ADVANCED GUIDED VEHICLES

ASPECTS OF THE OXFORD AGV PROJECT

WORLD SCIENTIFIC SERIES IN ROBOTICS AND AUTOMATED SYSTEMS

Vol. 1: Genetic Algorithms and Robotics — A Heuristic Strategy for Optimization (*Y Davidor*)

Vol. 2: Parallel Computation Systems for Robotics: Algorithms and Architectures (*Eds. A Bejczy and A Fijany*)

Vol. 3: Intelligent Robotic Planning Systems (*P C-Y Sheu*)

Vol. 4: Computer Vision, Models and Inspection (*A D Marshall and R R Martin*)

Vol. 5: Advanced Tactile Sensing for Robotics (*Ed. H R Nicholls*)

Vol. 6: Intelligent Control: Aspects of Fuzzy Logic and Neural Nets (*C J Harris, C G Moore and M Brown*)

Vol. 7: Visual Servoing (*Ed. K Hashimoto*)

Vol. 8: Modelling and Simulation of Robot Manipulators: A Parallel Processing Approach (*A Y Zomaya*)

Vol. 9: Advanced Guided Vehicles – Aspects of the Oxford AGV Project (*S Cameron and P Probert*)

Vol. 10: Cellular Robotics and Micro Robotic Systems (*T Fukuda and T Ueyama*)

Vol. 11: Recent Trends in Mobile Robots (*Ed. Y F Zheng*)

World Scientific Series in Robotics and Automated Systems – Vol. 9

ADVANCED GUIDED VEHICLES

ASPECTS OF THE OXFORD AGV PROJECT

EDITORS

STEPHEN CAMERON
Computing Laboratory
Oxford University

PENELOPE PROBERT
Dept. of Engineering Science
Oxford University

World Scientific
Singapore • New Jersey • London • Hong Kong

Published by

World Scientific Publishing Co. Pte. Ltd.

P O Box 128, Farrer Road, Singapore 9128

USA office: Suite 1B, 1060 Main Street, River Edge, NJ 07661

UK office: 73 Lynton Mead, Totteridge, London N20 8DH

ADVANCED GUIDED VEHICLES
Aspects of the Oxford AGV Project

ISBN 981-02-1393-X

Printed in Singapore by Utopia Press.

Preface

Why should I recommend anybody to read, let alone buy, a book that summarises a project carried out over the past six years at Oxford University? Especially so, as I am normally sharply critical of the genre, whether as a research engineer or as editor of book series and journals. My criticisms tend to be of two kinds. First, many books about projects swamp the reader with implementation detail that is often tedious and unilluminating. It is often hard to see whatever wood there may be for the detail of the trees. Worse, I mostly feel that the authors of such books are not themselves capable of seeing what are the principles, and what is the irrelevant detail. Second, some books report projects that are limited in scope and whose results are never exploited in practice. They are, in fact, forgettable. So, to sharpen up my initial question, why does the project reported in this book escape my two criticisms? There are, I believe two reasons:

First, because autonomous (or self) guided vehicles (AGVs from now on) are an important component of modern materials handling technology and automation. The Oxford project was based on a commercially fielded vehicle, marketed by Caterpillar, and was born of industrial needs to enhance the capabilities of that system. The vast majority of automated vehicles currently deployed in manufacturing follow preset paths indicated by wires, magnets, or strips. Their advantage is accuracy, but their disadvantage is inflexibility. They are hard to re-route as manufacturing requirements change. Even when they remain largely unchanged, there is a need for endpoint manoeuvering that defeats wire guidance. This has led materials handling companies to explore adding sensors to their vehicles to enable them to move a limited distance off wire. The GEC/Caterpillar vehicle represents a break with wire guidance, and is one of the first truly free ranging vehicles available commercially. Several hundred have been installed in manufacturing plants world wide. Now new applications in civil construction, dock automation, marine automation, agriculture, mining, and military systems are opening up. It is hardly surprising, in light of the industrial and military significance of AGVs, that university research groups have, particularly over the past five years, begun to build and experiment with them. However, the overwhelming majority of the work has been, and seems destined to remain, purely academic, especially as hardly any has been carried out in collaboration with industry. The Oxford project avoids my second criticism, however, because it has been carried out in industry from the start, was inspired by industrial problems, and its results have ranged from the normal academic output of papers, theses, and books to fielded software and hardware.

Second, because the design and implementation of AGVs constitutes a paradigm of the important engineering challenge of building intelligent sensor-based systems. The Oxford AGV project has explored many different sensors including sonar, vision, a structured light range sensor, and a proximity sensor. Sensor data fusion is a key challenge of the work, and this in turn requires the sensors to be characterised formally so that results from several can be combined. An intelligent sensor-based system needs to plan, and needs to do so at several different levels. The Oxford project has concentrated on navigation, local path planning, and on trajectory planning. Finally, the control of such systems presents a formidable challenge: the controller needs to be *real-time*, to enable it to deal with obstacles; *distributed*, to increase the bandwidth and to provide for graceful degradation, hence increased safety, since AGVs typically operate near people; and deal with intrinsically *uncertain* sensory information. The book brings together the ideas from the tens of research papers, several DPhil theses, and two books produced by the group over the past six years. The emphasis is on principles, not implementation details, in particular how the parts fit together to make the working whole. In this way, the book also escapes the first of my criticisms!

Inevitably, a project such as that reported here is a major collaborative effort. The project was conceived by Stephen Cameron, Hugh Durrant-Whyte, Penny Probert, and me. Project funds have supported a number of talented researchers: Alan Brown, Huosheng Hu, Nick Pears, Barry Steer, Bill Triggs, and Han Wang. We have been fortunate also to attract some very fine DPhil students: Martin Adams, Alex Cox, Will Dickson, Tony Hague, John Leonard, James Manyika. Chris Pearson, Jian Peng, Bobby Rao, and Ian Reid have all been particularly involved with this work. As noted above, the project has involved close collaboration with industry, from its earliest conception to its exploitation. Initially, we collaborated with the FAST division of GEC Electrical Products, particularly Malcolm Roberts, Mike Robbins, and Alistair Kinross. Alan Davies was influential in the formation of the project. Subsequently Malcolm Roberts started a new company Guidance and Control Systems (GCS) with whom we have had a particularly fruitful relationship. GCS now license the LICA board to be described in Chapter 3. As well, BP funded a extra-mural research award that brought Hugh Durrant-Whyte to Oxford, and subsequently endowed my chair in Information Engineering. Since at the outset of the project we did not believe (and we were in retrospect correct!) that vision was not sufficiently mature, we collaborated with NEL, particularly Robert Rixon and Graham Reid, on the development of a triangulation laser range sensor, that has subsequently been sold commercially. The two dimensional opto-electronic scanner was developed in collaboration with Colin Morgan of Oxford Sensor Technology, and Tony Williams (of Thorn-EMI and MJV Ltd.) provided substantial support. Last, but crucial to the success of the project, has been the work of the ACME (Applications of Computers to Manufacturing Engineering) division of the Science and Engineering Research Council. We have had immense support and encouragement from Bill Hillier, Guy Richards, Jacqui Williams, and especially Dave Mullins. ACME's consultants, Alan Pugh and Graham Parker, have consistently asked tough questions and provided sound advice.

Working on the project has been very satisfying, for what engineer does not hope to develop new ideas and then see them being used in practice. It has also been great fun.

Michael Brady
Oxford
April 1994

Contents

II SENSING AND NAVIGATION

III MODEL BASED PLANNING

Chapter 1

Introduction

Towards the end of 1988 three of the stronger members of the Robotics Research Group struggled down the narrow stairs to our mobile robot laboratory under the weight of our first robotic vehicle. The advent of this vehicle, known as '*the* AGV' or sometimes as 'the Turtle', marked the start in earnest of the Oxford AGV project, which had as its broad aim the investigation of robotic vehicle technology for industrial applications. This book describes some of the achievements of the first five years of the project, especially in sensing and planning. Although most of the material presented is at the frontiers of international research we have aimed to make the book comprehensible to a broad audience through including an amount of more general material and two tutorial chapters. The book is appropriate for courses at graduate and final year undergraduate level, as well as for workers in the field.

In such a book it is possible to include only a fraction of the work in the project, and we have concentrated on the areas which have been developed especially for this vehicle: an architecture and algorithms to provide a reactive sensing-planning capability; sensors and sensor processing techniques; and finally more centralised global planning of paths and tasks to ensure safety and consistency. We include review material and research results to illustrate the potential of the work and its achievements. Altogether we believe that this book gives a sound balance and overview of the major issues and achievements in mobile robotics research, both in Oxford and elsewhere in the world.

The *autonomous* guided vehicle (AGV) was a donation from the FAST Division of GEC Ltd., who were based in Rugby (UK). The original inspiration for the project is a commercial range of *automated* guided vehicles, which were developed jointly by GEC-FAST and Caterpillar Ltd. for factory and warehousing applications. The Oxford AGV is a small version (0.6m $\times$ 1.1m base, 0.6m high) of the commercial vehicles, but with the same kinematics (i.e., it steers the same way) and supplied with an identical control computer. Unlike most automated guided vehicles the GEC/Caterpillar system is *free ranging*—that is, the vehicles do not have to follow fixed tracks or routes. The commercial system is an example of the state of the art, but has two important limitations: it cannot sense its environment, other than with the use of touch switches; and it cannot react intelligently when problems occur (such as obstacles getting in its way). Such an arrangement was acceptable for that generation of factory vehicles, but our work is aimed at developing the techniques for building more flexible and intelligent systems.

In the course of our work we developed novel devices and algorithms, and tested them using computer simulations, prototype devices, and on the AGV itself. This work was sponsored by a number of organisations, but most notably by the British government through the ACME directorate of the Science and Engineering Research Council. Although only a part of the work being undertaken by the Robotics Research Group, it did force us to confront head on the problems of integrating often disparate systems and ideas. Work on *the* AGV has spawned other projects within our group based on other guided vehicles and other devices, many of which are on-going; the purpose of this book is to bring together some of the key results of the AGV project to date in a coherent and comprehensible form.

One early decision that we had to make was whether to do most of the computer processing off the vehicle, or on the vehicle. On-vehicle processing makes more sense if you are trying to build truly autonomous devices, whereas for many industrial vehicle applications you need to coordinate many vehicles, and so would prefer the vehicles to be *tasked* by a central control computer. Thus it was clear that at least some of the processing of a vehicle system would occur off-vehicle. However in a multiple vehicle system bandwidth limitations tend to restrict the amount of information passing between each vehicle and any central controller, and hence limits the dynamic planning and scheduling which can be performed. Our work has concentrated on a mixture of these two extreme views, with the provision of some local autonomy for each vehicle, so that each makes local decisions without reference to a central computer, and a central computer that provides a list of tasks to each vehicle in some suitable high-level form. Producing such a *distributed* control architecture is a hard problem in itself, and so this added a third major theme of *architectures* to the AGV project (and thus to this book).

Structure of this Book

The book is essentially organised in a 'bottom-up' manner, with the early chapters describing material which is close to the vehicle itself, and the later chapters describing material which is not clearly tied to driving a particular vehicle. As a consequence the work is not presented in any form of chronological order.

The book is in three sections which describe three major aspects of mobile robotics:

architectures, sensing and planning. The multi-sensory and multi-tasking needs of robotics impose considerable demands on architecture design. There are a number of requirements, sometimes conflicting, such as fast reactivity for local planning, data representations which can be updated dynamically to handle complex environments, and a means of maintaining the consistency of internal and environmental state even when there are a number of sensors and different levels of planning. Two possible architectures are described in the book, the first, in chapter 3 which emphasises a tight loop between sensing and planning for reactive control and a second, in chapter 13 which emphasises data representation for more considered planning. Essentially these two architectures are examples along a continuum of architectures, with the first emphasising autonomy and local decision making, and the second emphasising centralised control with distributed computing elements.

The first section of the book, *Real time architectures for sensing and planning*, starts with a chapter describing the control system and architecture supplied in the commercial factory vehicles and on the *AGV*. The original controller includes low level functions—such as a communications module to the central *landbase* computer, a motor control module and an error identification module—realised through a number of 8-bit microprocessors co-ordinated by a programmable logic controller. An innovative and particularly successful feature is the self localisation capability based on a laser scanner which detects large bar codes placed at intervals around the workplace.

Chapter 2 describes the vehicle as we acquired it, but more powerful computing is required to increase local functionality. The next two chapters describes the replacement of this controller with a transputer based architecture for reactive sensing and planning. The architecture also provides a simple dynamic planning capability, replacing some of the functionality of the central landbase. Chapter 3 starts with a review of architectures used in major robotics projects throughout the world. Following this, it develops functionality based on a loose intercommunicating network of locally intelligent control agents which draws from behavioural as well as more conventional task directed designs. Local message based communications ensure reactivity and avoid the bottleneck of a centralised data structure for sharing information. The architecture is realised in the transputer which uses links for local communications in an essentially synchronous message based environment. The chapter describes the development of modules for functions such as sensing, vehicle control and obstacle avoidance.

Chapter 4 describes the application of this architecture in global and local factory planning. Early work on the vehicle achieved a simple planning and obstacle avoidance strategy using a sonar array to provide environmental information. However this left many questions unanswered: for example how to handle uncertain information, when to stop sensing and decide on the next action, how to make use of earlier information. These have been addressed in this chapter through algorithms based on Bayes statistics in a decision theoretic context to maximise the use of uncertain information. The algorithms and data structures have been developed for computational efficiency to provide a real time facility.

Reliable sensing underpins reactive planning and the next section of the book, *Sensing and Navigation*, describes sensors for local reactive planning and for navigation in a global plan. The main requirement is information on the distance or *range* of objects near to the vehicle in the environment. Suitable sensors must be chosen through considering factors such as cost, environment (for example, lighting and the types of surfaces expected) data representation, and acquisition time. Chapter 5 (a tutorial chapter) describes how suitable sensors might be chosen in the light of these factors, and especially on the use of low cost range sensors for directing fast planning. These sensors are *active*, in that they transmit radiation into the environment and determine range through one of several techniques from detecting the received beam. The chapter describes the application of technologies based on ultrasound and electromagnetic radiation, presenting performance information which can be used to select factors such as wavelength and radiation mode. Two popular methods—sonar and optoelectronic line of sight sensing—are described in some detail, presenting work in Oxford and elsewhere.

A second chapter on sensing (Chapter 6) describes another popular optical technique: ranging through triangulation. Triangulation sensors are constructed to project a beam at an angle so that the lateral offset at which the reflected beam returns to the sensor depends on how far away is the object reflecting it. Conventional triangulation ranging has been used to produce three dimensional range images by projecting a laser stripe into the environment, and with the range detected by measuring the offset of the returning beam on a CCD camera. The disadvantage of this method is the time that it takes to acquire a single range image. To provide rapid ranging we investigated a different device for detecting the returned radiation, namely the lateral effect photodiode. The chapter provides a thorough discussion of the choice of triangulation method, geometry, detector, and source. It present results from a high bandwidth sensor developed in Oxford.

Sonar is probably the most popular low cost ranging sensor used in robotics. A major problem associated with sonar is the large beamwidth which makes angular resolution poor. The twin-sonar tracking head described in Chapter 7 overcomes this problem by using a differential technique. As well as sensor design, an important result of this and previous work is the choice of environmental representation, which uses common features to which sonar is highly sensitive. The chapter is illustrative of the large body of work in sonar modelling and localisation which has taken place in Oxford.

Probably the commonest method of sensing uses passive vision (i.e., using images taken by a camera without special illumination of the scene). Hardware is now becoming fast enough to offer the possibility of using passive vision as a real time sensor, although there are still considerable difficulties in image interpretation in general environments. Developments in the acquisition of structure from motion has lead to new classes of algorithm which are more amenable to real time performance than conventional techniques of stereo matching. The possibility of using vision to direct real time planning and navigation is illustrated in Chapter 8, which discusses the *Droid* vision architecture developed in conjunction with Plessey and British Aerospace.

The final chapter in this section (Chapter 9) looks towards planning. The context of the work described here was warehousing, with a requirement to to move pallets whose exact location is not known. To add to the complexity the pallets may be of several types, thus giving a problem in recognition as well as in position determination. A generic pallet model is introduced and used to direct model based recognition algorithms. The sensor modality is important; a three dimensional image is needed but passive vision is not robust enough to be used outdoors. Range data is more suitable and this chapter links to Chapter 6 through its description of the triangulation rangefinder developed by Oxford and the National Engineering Laboratories (East Kilbride) for this task. The range data is used to set up a specially constructed constraint network within a computer, and this is solved using a novel approach to determine the position and type of each pallet. This method has been tested using hardware installed on the AGV itself.

Whereas the first two sections of the book are concerned more with the development of the pieces that allow a vehicle to be built and to sense its environment, the last section of the book—*Model Based Planning*—is concerned with planning what the vehicle should do. A major part of the planning problem is that of finding paths for the vehicle that avoid collisions. A large body of work has been done in this area, and Chapter 10 reviews the relevant material in a tutorial manner.

Many of the path planners developed in research laboratories are either impractical or inappropriate for the AGV, which has a wheel arrangement that is car-like and *non-holonomic.* (Non-holonomic is a term with a rather technical meaning, but the essential problem that arises is that our AGV cannot turn on the spot.) We have developed a path planner for the AGV (and other vehicles) that is both effective and practical, and this is described in Chapter 11. The path planner is based on the *potential field* paradigm, which uses the idea of attracting and repulsing electrostatic charges to direct the vehicle. Our technique differs from that used in other implementations in using the potential field to directly steer the vehicle, and in being able to use higher-level structural information about the vehicle's environment to speed up the search for a path.

Although path planning is important it is only a means to an end; most vehicle tasks are naturally expressed in terms of what the vehicle is to achieve, rather than in terms of paths to be planned. In Chapter 12 we introduce a *task* planner that is capable of taking a task specification and generating a sequence of motions to accomplish the task. The planner contains a number of (generic) rules that encode the ways that the vehicle can be used, and some further (task specific) rules that encode how the task can be achieved. The planner applies the rules to look for a solution for the task, changing its model of the world as it goes. What makes this work different from conventional AI task planning is the use of a number of *geometric experts*—such as path planners—that are capable of handling realistic geometry. The task planner was developed to handle the problem of manipulating palleted goods, but designed in such a way to make it easy to apply the planner to other problems.

Finally, we consider the issues of using and generating internal representations or *models* within the centralised planner. Many research-oriented planners finesse this problem by using simple models, whereas for industrial use we need to be able to cope with models that are large and complex. Chapter 13 starts by describing the *world model*, a software architecture for a modelling system specifically designed for these applications. The world model has a hierarchical structure which leads to an efficient implementation, even with models composed of many thousands of elements, and is capable of storing information which is not of a purely geometric form (such as the sensor signature of interesting features that the vehicle has discovered). We also describe in this chapter ways of generating sensible 'corridors' automatically from the geometric data, and show that this information can be used to great effect with the path planner described in Chapter 11.

Part I

REAL TIME ARCHITECTURES FOR SENSING AND PLANNING

Chapter 2

The Oxford Project and the GEC AGV

Huosheng Hu and Penelope Probert

The mobile robot project in Oxford started in collaboration with GEC FAST division. The AGV, provided by GEC Fast, has lived in our AGV Laboratory since August 1988. It is a small (0.6m × 1.1m) research version of the commercially-available, free-range AGV designed for transporting materials around factories. In the commercial version, several vehicles work together in the factory, controlled by a central *landbase* computer as shown in figure 2.1 [75].

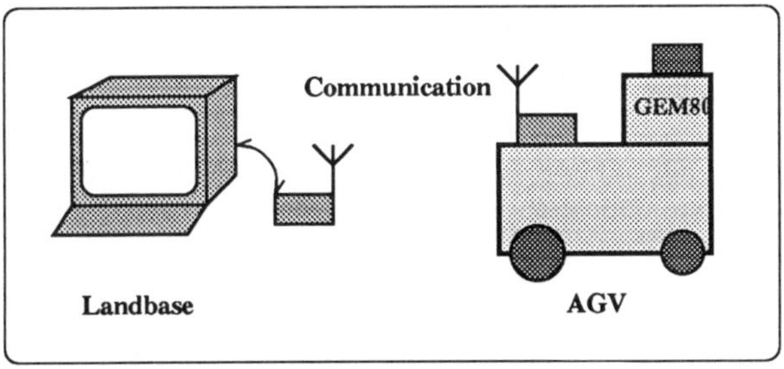

Figure 2.1: Landbase controlling the AGV

Mechanically the AGV is a three wheeled vehicle, steered by a single front wheel. Several other vehicles are also used in Oxford: in particular the Robosoft *robuter* (a four wheeled platform), a small circular platform and several modular navigation platforms.

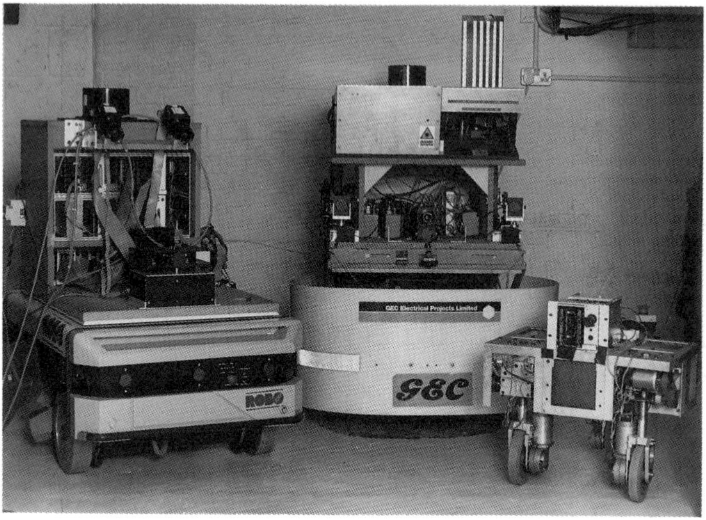

Figure 2.2: Some of the Oxford vehicles

The work described in this book was developed largely around the AGV vehicle. Unlike most automated factory vehicles, the AGV is free ranging - there is no hard wired control through buried wires or painted lines. For localisation the factory is equipped with a number of barcodes, which are at known positions, and each vehicle uses a rotating laser scanner to read the barcodes and determine its position (as explained in §2.1.2).

The landbase, a microcomputer system, holds details of the factory, in terms of allowed routes (including speed limits and width restrictions) and the position of barcodes. During AGV operation its tasks are as follows:

- To keep track of and display AGV positions, headings, and operational parameters and alarms.
- To pass path information to each AGV and receive position and status information. Communication is through a multidrop serial protocol, called the Extended Simple Protocol (ESP).
- To schedule multiple AGVs

Continuous bi-directional communications between the landbase and the AGV is essential to allow the landbase to dynamically control the robot motion and allow the AGV to report the status of its operation. To ensure there are no collisions, a simple scheduling plan divides the factory up into a number of sectors, and ensures that only one vehicle is in a sector at a time.

A central global controller has the advantage of simplicity, but causes a bottleneck in communications and processing. The Oxford project emphasises local decisions and intelligence to make the AGVs truly autonomous, able to make their own decisions based on sensing actions.

2.1 On-board GEM80 Controller

In the commercial version, the landbase performs higher level functions, but each vehicle has on board a local controller, co-ordinated by a GEM80 programmable logic controller [74]. The GEM80 controller is supplied complete with power supply, microprocessors(Intel 8085, 8088, 8086), memory and space for 8 small I/O modules which can be configured to suit the industrial applications. It has a tightly coupled multiprocessor architecture in which all processors share a common memory. As common in that generation of programmable logic controllers, it is programmed using a ladder diagram which offers a development environment for different applications. The GEM80 is a cycle controller. Its cycle time can be changed according to different applications. For the AGV, it has been chosen to be 80 ms. Figure 2.3 shows the configuration for the control of the AGV and a cycle of the operation.

We describe the GEM80 controller in some detail in the rest of this chapter, since we still use much of its functionality for low level control (particularly localisation and guidance). As we describe in chapter 2. the next stage of the project replaces the controller by a distributed transputer architecture.

The GEM80 control task is functionally decomposed into different function modules, shown in figure 2.4. The architecture is hierarchical, combining three main functions: message handling, system management, task control [76].

2.1.1 Control Strategy

The control of the AGV is based upon two feedback loops as shown in Figure 2.5. The inner loop consists of two parallel feedback loops for motion and steering control. Two traditional analogue PID controllers are employed to ensure stability in speed and steering angle. The outer loop based on the GEM80 is used for position control of the robot. The odometry and bar codes information are fed back to a navigation system inside the GEM80. A guidance system tracks the position demand to follow the desired path issued from the landbase or preset in the memory. Based on an error-feedback control principle, both loops automatically respond to unforeseen system change or to unanticipated disturbances. This control system has no capabilities regarding obstacle avoidance and path planning. If an obstacle is placed in its way, the AGV can only use its physical plastic bumper to stop and wait for the obstacle to be removed.

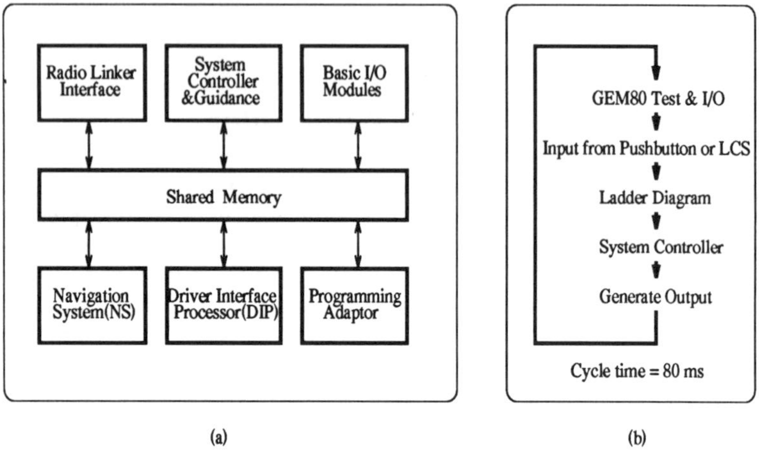

Figure 2.3: The GEM80 Controller for the AGV

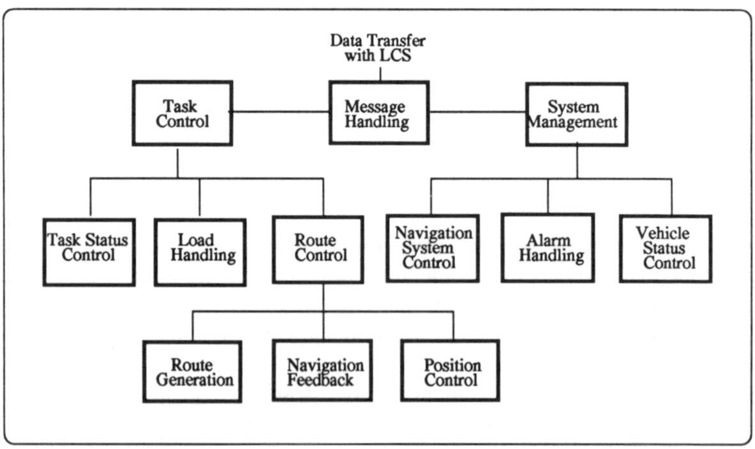

Figure 2.4: Block diagram of the modular structure

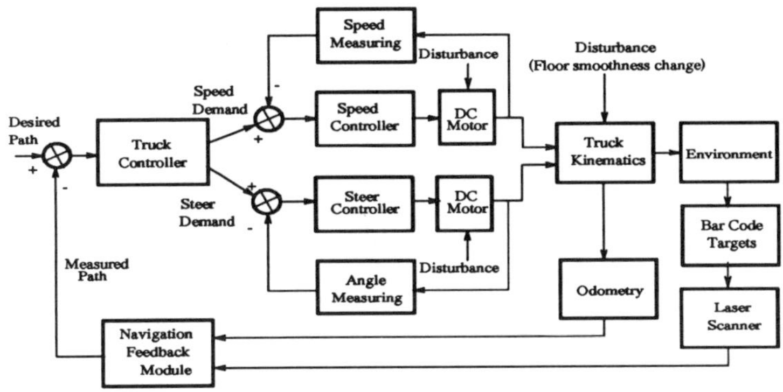

Figure 2.5: Two feedback loops for the AGV

2.1.2 Laser System: Localisation

One of the key requirements in AGV control is localisation; knowing where the vehicle is. The position of the AGV is determined from data provided by a laser scanner situated on the top of the vehicle and by odometry. Light from a laser diode (class 3b) is scanned in azimuth at a constant speed of rotation at two revolutions per second. Passive retroreflective targets, in the form of bar codes, are positioned around the working area; the reflected light from these targets is received by a photodetector in the scanner unit. Each target is a unique 5 bit code and the absolute position of each target is known.

Prediction of position from the bar code data is accurate to within about 1cm, easily adequate for the factory environment. However, bar codes may sometimes be hidden from the vehicle's view. Therefore the localisation module combines data from the scanner with shaft encoder information, using a Kalman filter (Figure 2.6).

2.1.3 Guidance System

Figure 2.7 shows a block diagram of the guidance system. The message handling process is prepared to receive one task and its corresponding route description file (up to 59 steps) each cycle time from the landbase (a new task will not be accepted by the system until the present task has been completed). A curve fitting algorithm generates a smooth trajectory of curves and straight lines, and reduces these to incremental reference vectors, described by pathwidth, position, heading and speed. These vectors are stored and used together with feedback information, to generate speed and steer demands to the vehicle controller each cycle period.

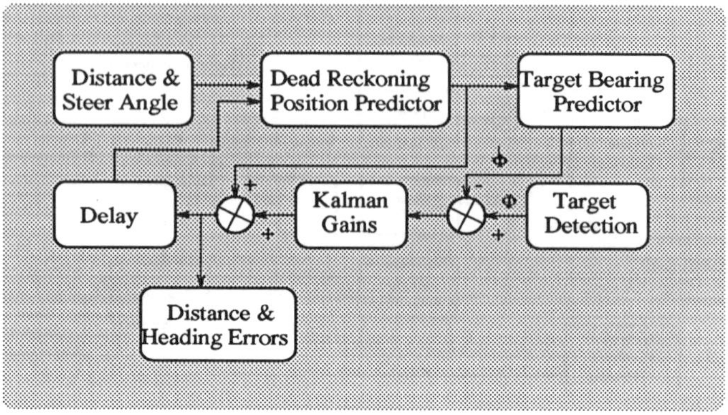

Figure 2.6: Block diagram of the position measuring algorithm

2.2 Overview of the Control System for the AGV

Finally Figure 2.8 presents an overview of the AGV control system. All systems are co-ordinated through the GEM80 ladder diagram.

- A rotating infrared laser sensor reads the bar codes on the wall and sends the measured code information back to the feedback module.
- The feedback module consists of three processors (SIP, NS, DIP). The SIP processes data from a laser sensor and transmits target code and angle information (in two words) to the navigation system (NS) whenever a valid target is seen. The DIP, a modified version of the GEM80 basic I/O, collects odometry data and sends it to the NS. Using both sets of data, the NS runs a Kalman filter to locate the AGV position and stores the results in a navigation data file.
- The guidance system handles only one task and its corresponding route description file at each time. It uses a curve fitting algorithm to generate a smooth trajectory as a position demand. It uses this together with position feedback information to generate the speed and steer demands which are sent to the vehicle controller through the DIP.
- The vehicle controller (VC), which is based on an Intel 8088 processor, receives the motion demands, either from the DIP in a automatic mode or from a joystick in a manual mode, to control the drive speed and steer angle of the front wheel. It also sends back the odometry information to the navigation system via the DIP.
- Two drive units, each with two feedback control loops(current and speed), drive two DC motors. The conventional analogue PID controllers are used to maintain

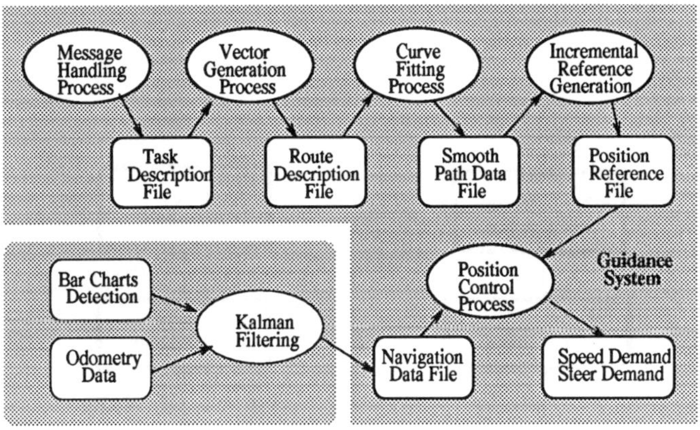

Figure 2.7: Block diagram of the guidance system

the stable speed and steering angle.

2.3 Functionality Addressed in the Oxford AGV Project

The first stage of the Oxford project added sensors to allow the vehicle to make local decisions if its current plan became unachievable: eg because the path was blocked. A new architecture has now been developed to replace the GEM80 and integrate higher and low level planning. Further work is in sensor integration and in geometric methods for higher level planning. These areas are described in later chapters of this book.

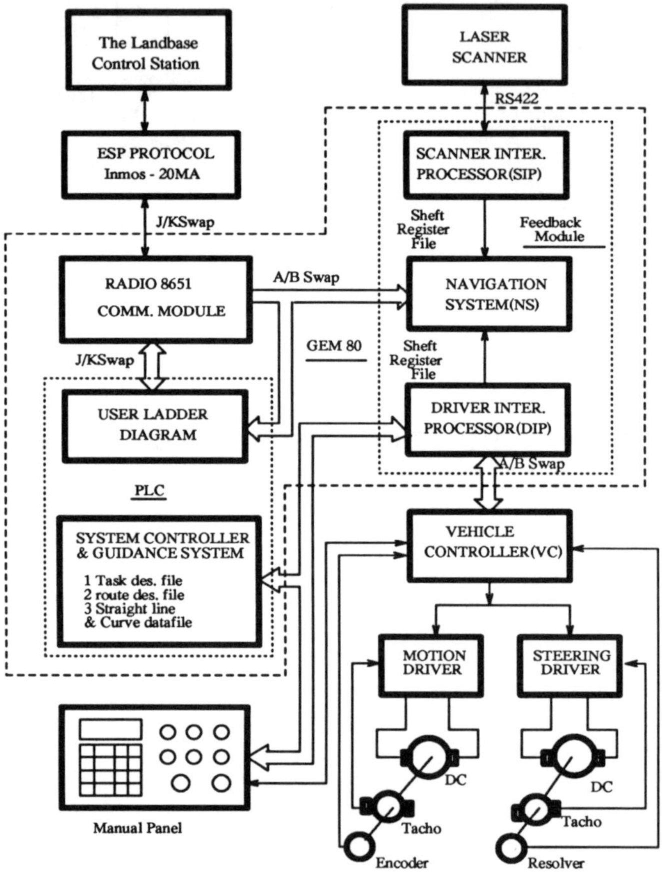

Figure 2.8: Block diagram of control system configuration

Chapter 3

Sensor-based Control Architecture

Huosheng Hu

In order to achieve a high degree of autonomy and robustness, a mobile robot should be able to sense its surroundings, build or update maps, plan and execute actions, and adapt its behaviour to environmental changes. To implement these tasks demands a sensor-based control architecture to provide for unified action in real time. In this chapter, we describe the system design of a real-time, sensor-based control architecture for our mobile robot: the distributed real time architecture (DRTA). A distributed design for sensing and control, as well as its implementation on a network of transputers, is presented.

3.1 Introduction

Sensor-based control provides the capabilities for a mobile robot to interact intelligently with a dynamic environment while performing specified tasks. To support such a sensor-based control strategy, many distributed architectures have been developed recently in mobile robotics, ranging from very specific ones for a particular mobile robot to very general ones that are intended to be suitable for a wide variety of applications. However, no specific architectural approach can be identified as best so far [86]. Therefore, the design of a suitable, robust and reactive architecture remains a key challenge in robotics.

In general, as the capability of a mobile robot increases, the number of sensors, motors and processing components that must be integrated and coordinated also increases. Therefore, a distributed computing architecture offers a number of advantages to handle the significant design and implementation complexity inherent in sophisticated mobile robot systems. It is often cheaper and more resilient than the alternative uniprocessor designs. Also, multiple, possibly redundant, processors provide the opportunity to take advantage of parallelism for improved throughput and fault tolerance. The modularisation necessary for a distributed implementation often simplifies the difficulties of complex hardware and software design as well as debugging. The most important aspect of a system design is the need to meet real-time constraints, which is considerably easier given the concurrency inherent to distributed computing solutions. Fortunately, transputers provide more efficient ways to build such distributed architectures.

3.2 Previous Work

During the last two decades, the research community in mobile robotics has paid a lot of attention to the development of different control architectures. Two principle designs have been adopted. One is called the functional, or horizontal, decomposition; the other is the behavioural, or vertical, decomposition [33].

Functional decomposition is a classical top-down approach to building systems. In this approach, the entire control task of a mobile robot is divided into subtasks which are then implemented by separate modules. These functional modules form a chain through which information flows from the robot's environment, via the sensors, through the robot and back to the environment via actuators, closing the feedback loop, as shown in Figure 3.1.

Most previous mobile robots have been based on this approach. Both hierarchical and blackboard architectures have been used, such as the Shakey robot at SRI [149], the Hilare robot at LAAS [77],[151], the Alvin robot at HAIC [57], the Ground Surveillance Robot at RII [85], the CMU Rover [62], the IMP robot [55], and the Navlab robot [78]. The hierarchical architecture is a highly deterministic and allows for the coordination of problem solving strategies [156]. It naturally decomposes a large complex system into small modules for relative ease of implementation. More details can be found in [4], [5], [92]. The blackboard architecture includes a central database, a pool of knowledge-intensive modules, and a system manager which synchronises the modules. It provides a very powerful mechanism through which to fuse various types of sensor data to realise a loosely coupled computer network. However, both a hierarchical and a blackboard architecture have limitations. A pure hierarchy precludes interaction with low-level sensing and action processes at higher levels and tends to result in very structured decomposition of problems leading to overall longer processing times and poor use of sensory information [155]. Its strictly structured decomposition makes it difficult to deal with uncertainty. This sequential structure requires an instance of each module to be built in order to run the robot at all. The blackboard structure has the problem that the shared database may cause bottlenecks in access, imposing a limit on the bandwidth of the whole system.

In contrast, behavioural decomposition is a bottom-up approach to building a system. A behaviour encapsulates the perception, explore, avoidance, planning, and

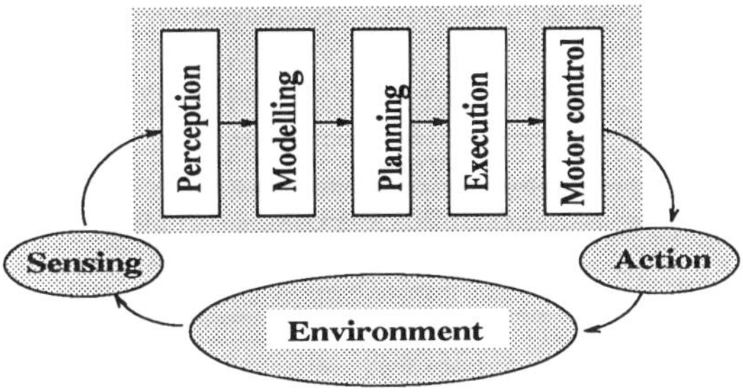

Figure 3.1: Horizontal decomposition of the control system

task execution capabilities necessary to achieve one specific aspect of robot control. Therefore, each of them is capable of producing meaningful action which in turn can be composed to form levels of competence [36]. In other words, each of them realises an individual connection between some kind of sensor data and actuation. The system is built step by step from a very low level, say from locomotion, to obstacle avoidance, to wandering. Successive levels can be added incrementally in order to enhance the functionality of the robot, as shown in Figure 3.2.

This design method has grown in popularity recently. Examples include Brooks' subsumption architecture [35], Anderson and Donath's emergent behaviour [59], Kaelbling's intelligent reactive system [112], and Arkin's schema based architecture [9]. In the subsumption architecture, control is distributed among those task-achieving behaviours that operate asynchronously. Lower layers can subsume the operation of the higher ones when necessary, only one layer at a time actually controlling the robot. Since each layer is to achieve a single task, it requires only the limited information that is useful for its operation. It is claimed that the control system can respond rapidly to dynamic changes in the environment without the delays imposed by sensor fusion. But the implementation of higher layers of competence still poses a problem. More careful initial design in specifying the communications and modularity is required. Moreover, the higher levels often rely on the internal structure of lower levels, thus sacrificing modularity. In Kaelbling's architecture, control is broken down into levels that are similar to Brooks's. The key difference is that each level consists of a set of parallel decision modules whose output are fed to a mediation unit. Arkin develops motor and perceptual schemas each of which is an individual computing agent. His approach is strongly influenced by cognitive psychology and neurophysiological studies. Although it shares the viewpoint of the necessity of behavioural expression and the importance of the immediacy of sensory information, Arkin's approach differs from Brooks' subsumption architecture by avoiding layering entirely; instead setting up a dynamic network of schemas based on the current goals of the robot. Control of the robot is achieved through schemas which implement basic

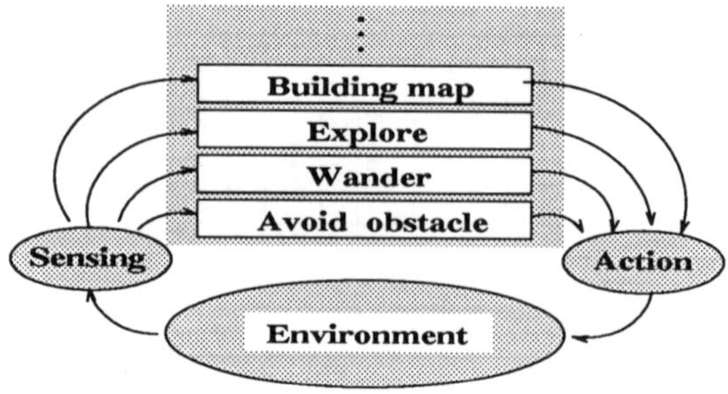

Figure 3.2: Vertical decomposition of the control system

behaviours. To ensure a high degree of concurrency, schemas communicate through a blackboard. The simulation result has shown the feasibility of this approach, but it is necessary to find better ways for real time applications [131].

We believe that the control task associated with a mobile robot is so complicated that we cannot strictly follow one decomposition scheme while completely ignoring the other. In fact, every approach has both benefits and limitations. A compromise, or combination, of both schemes may be a better choice to build more robust, flexible, and high bandwidth architectures for mobile robots.

The control architecture we designed and implemented for the AGV is a hybrid structure which merges positive features of both the hierarchical and subsumption approaches. It is called a distributed real-time architecture (DRTA), and is based on a transputer network [99]. A similar hybrid approach has recently been adopted in other mobile robot projects. For example, Badreddin proposes a *recursive nested behaviour control* structure in [10] where the behaviour levels implement feedback and feedforward compensations. Knieriemen and Puttkamer suggest an *orthogonal control* structure with cascaded control loops, combining aspects of hierarchical and behavioural structures in [115]. Mitchell presents a Theo-Agent architecture which combines a stimulus-response subsystem for rapid reaction, with a search-based planner for handling unanticipated situations [144]. The key achievement of the hybrid approach is in merging the positive features of the strictly functional and behavioural approaches.

3.3 Distributed Design

The original AGV controller was based on a GEM-80 controller, and has certain limitations for advanced research work. Therefore the control architecture of our mobile robot has been re-designed to satisfy the following criteria:

- Robustness – to cope with the inaccuracies and uncertainties of sensor data as well as with hardware errors like communication failures.

- Modularity – to distribute sensing and control among different modules for task-achieving behaviours.
- Expandability – to build up the entire system stepwise from low-level behaviours like obstacle avoidance to high-level behaviours like path planning.
- High bandwidth – to perform real-time planning and control tasks.
- High intelligence – to cope with uncertainty in a dynamically changing environment, including:
  - the ability to generate a solution in unexpected situations.
  - the ability to learn something new from its travels.
  - the ability to communicate with other robots and humans.

Figure 3.3 shows a block diagram of the control architecture. It consists of a distributed community of sensing, action and reasoning nodes. Each of them has sufficient expertise to achieve a specific subtask, following a hierarchical decomposition scheme. Few of them are composed of a single task-achieving behaviour as in a behavioural decomposition scheme. The key point is that the control task of the mobile robot is distributed among a set of behaviour experts that tightly couple sensing and action, and which are also loosely coupled to each other. In this way, particular sensor data can be used directly in a corresponding layered control task to form a task-achieving behaviour. This differs from the functional decomposition in which the combination of subtasks is a single processing chain. To function effectively, each layer requires a significant amount of self-knowledge to allow it to decide what information it can supply to others; how best to recover from local errors; as well as what to do if no information is sent from other layers.

Basically, there are two subsystems in our design: a layered perception system and a layered control system. The layered perception system is used for active sensor control and distributed sensor fusion to support a layered control strategy. The layers process raw sensor data from internal sensors (tachometer, encoder, resolver) and external sensors (sonar, vision) to build up models in a bottom-up manner. All the sensing layers operate independently and are loosely coupled. Communication between them is used to realise sensor data fusion. The design of the layered controller is based primarily on the observation that different response times are demanded by different tasks. The lower levels perform simple, more general tasks such as *smooth path guidance* and *avoiding obstacles* for fast reactivity. The higher levels perform more complex, situation specific tasks such as *path planning* and *monitoring*. All layers are intended to operated in parallel.

Each layer of our DRTA consists of a control node and a sensing node. Each sensing node delivers a representation, which, in the context of a given task, causes a corresponding control node to generate commands. Here, we try to avoid a centralised and complicated model for a dynamic environment since it needs more time to compute, thereby reducing the system's response speed. Instead, we are employing several simple models, each tuned to a different range and resolution of situations for

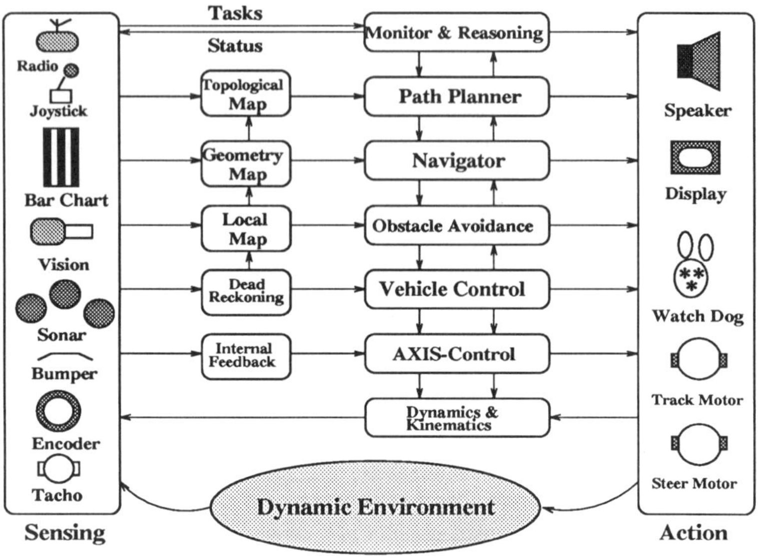

Figure 3.3: Block diagram of the control architecture

different tasks. In other words, this multi-model control architecture takes the real-time capability and the expected task-achieving behaviours into account. In more details:

- the top layer (Monitor & Reasoning) is used mainly for communication tasks which include receiving and interpreting mission tasks, $t_i \in \mathbf{T}$, from a planning centre and reporting on operation status. It also tells the operation status to humans through an audio system.
- The next layer (Path Planning[1]) works on a topological map to determine the subgoals to reach an assigned task. At this level of resolution, the world representation consists of a network of distinctive places such as rooms, corridors, junctions, etc. It forms a predefined graph with a cost function on each arc based on prior information, which is then updated according to sensor data. An optimal path, $\mathbf{n}_i \in \mathcal{N}$ between the start node $\mathbf{n}_s$ and the goal $\mathbf{n}_g$ can be planned based on this graph.
- The task of the third layer (Navigator) is mainly to update the robot position. A rotating infrared laser scanner situated on the top of the vehicle is used to read

[1]This is essentially an 'on-line' path planner, as opposed to the 'off-line' planning described in later chapters

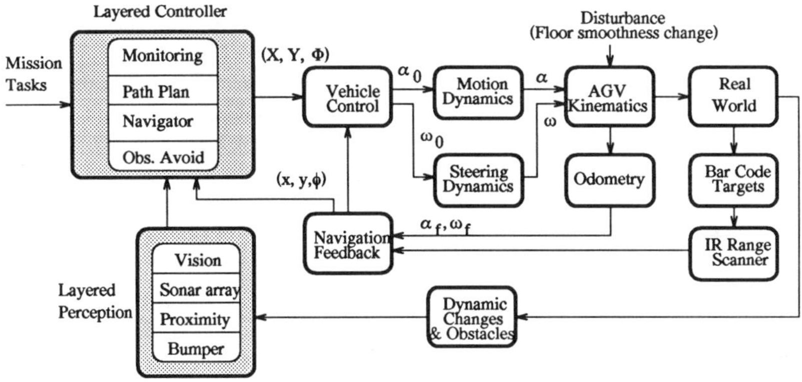

Figure 3.4: Control strategy for the AGV

retroreflective bar codes that are attached to the factory wall at known locations. The position measuring algorithm is capable of simultaneously accepting data provided by the laser scanner and by odometry. Since both set of data, the odometry data (α, ω) and the measured angle λ to each bar code, are noisy, a Kalman filter is used to estimate the position (x, y) and orientation θ of the robot in real time.

- The fourth layer (Obstacle Avoidance) uses sonar data to build a local map by linear approximations of range and azimuth information (r, α), and provides the capability to avoid obstacles which are unknown to (or ignored by) the Navigator.
- The vehicle control layer below obstacle avoidance plans a continuous-curvature trajectory and maintains the continuity of motion along this trajectory to achieve smooth guidance and stability of overall system.

Traditional methods in control stress the use of feedback. Feedback is naturally adopted in our layered design, and described in Figure 3.4 in the context of the control system design. Note that the *planning* problem is viewed from the perspective of a *control* problem in the design. To implement this layered control strategy, we describe a real time computer architecture in the next section.

3.4 Real-time Computer Architecture

The key feature of the DRTA is that all the layers in the system should operate in parallel and be loosely coupled to each other. This naturally demands a multiprocessor computer architecture since it is unlikely that all the robot's tasks, such as navigation, map building, planning, and control can be carried out concurrently by one processor. This multiprocessor architecture should be capable of real time, on board processing that is modular, reliable, and expandable. Also, low power consumption and the use of commercially available products are important features. However, multiprocessor control of a mobile robot is complicated and poses problems that don't exist in a uniprocessor system. Typically, the following factors must be taken into account in the design for a particular application:

- the degree of coupling.
- the configuration of processor interconnection.
- the method of deadlock detection and protection.

3.4.1 Conventional Approach

Multiprocessors are general-purpose, asynchronous parallel machines with multiple instruction-streams and multiple data streams [70],[101]. They can be classified as being *tightly coupled* or *loosely coupled.* Processors in a tightly coupled multiprocessor communicate through a shared memory, even though they may still have their own private memory. In such an architecture, there are two ways to access the shared memory: via a shared bus (Figure 3.5(a)) or via an interconnection network such as a crossbar (Figure 3.5(b)).

In contrast, loosely coupled multiprocessors have no globally shared memory. Instead, processors communicate by sending and receiving messages, which typically use interconnection networks with direct, point-to-point connections between processors, as shown in Figure 3.6. In this architecture, communication between processors can be time-multiplexed through a common bus; switched through a limited number of channels, as in a crossbar scheme; or implemented through dedicated links. A loosely coupled configuration is also called a *distributed architecture.* Klein and Wahawisan proposed a loosely coupled system for the control of the OSU Hexapod vehicle, using an inexpensive, high density, eight-channel parallel line unit [118]. They implement different configurations (tree, star, and loop) with a fully connected system, as shown in Figure 3.7.

To build these multiprocessor systems using conventional processors is very difficult in terms of the implementation of both the hardware and software. In the hardware implementation, communication between processors has to use methods such as shared memory or a common bus since conventional processors follow the *Von Neumann* architecture that is sequential in nature (the instructions and data share a common path and are fetched into a central processing unit (CPU) one by one). The lack of any special communication unit on-chip for communication with other processors makes the implementation complicated and the configuration inflexible. It also leads to some serious limitations such as bus contention, which arises when more than a pair of processors request communication in a shared bus system,

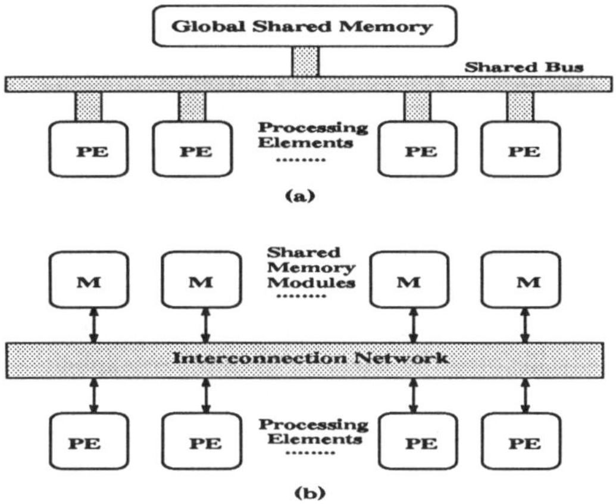

Figure 3.5: Tightly coupled multiprocessor

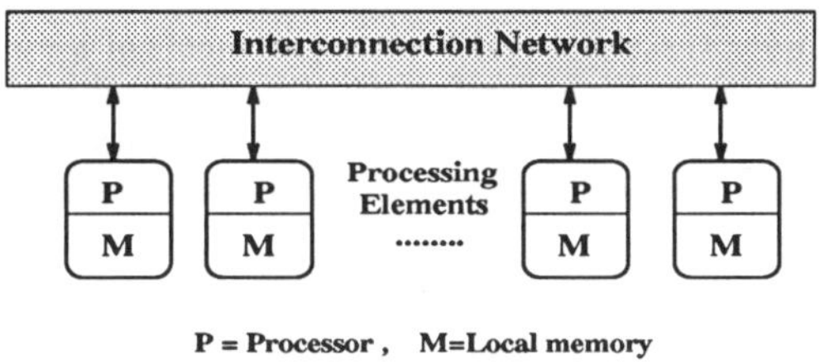

Figure 3.6: Loosely coupled multiprocessor

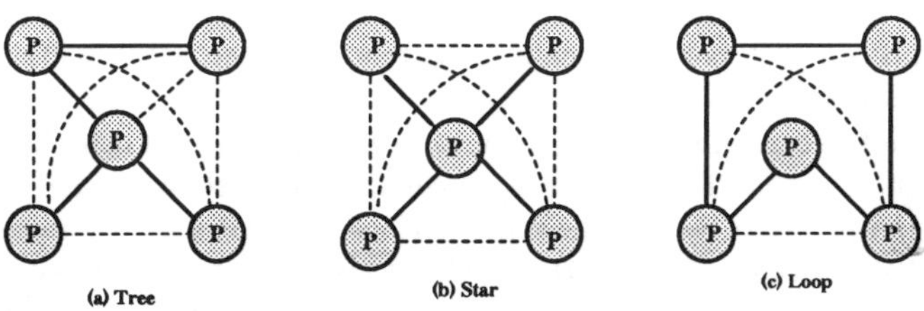

Figure 3.7: Different configurations with a fully connected system

or bottlenecks in access to the globally shared memory, or saturation of its bandwidth as the size of the system increases.

As regards software implementation, a major obstacle is the lack of convenient parallel programming languages. The languages currently available on multiprocessors are essentially existing sequential languages (C, Fortran, Pascal) with a few extensions for parallelism, such as Concurrent C [52], Parallel Fortran [6], and Concurrent Pascal [83]. As a result, difficulties exist for fully exploiting the potential parallelism provided by the hardware. This leads to systems that are complex and unportable. Also, parallel hardware may not be able to implement efficiently certain constructs of the original sequential languages. For this reason, the trend appears to be towards languages with explicit parallelism and partition, such as *Ada and Multilisp* which offer portability but still force the users to decompose explicitly the program into tasks and to control their synchronisation and communication [181]. Another major problem in concurrent programming is ***deadlock***, in which all affected process are suspended indefinitely waiting for one another. It is extremely unlikely that compilers will be intelligent enough to detect deadlock. It is thus necessary for designers to demonstrate that the possibility of deadlock has been removed from their software [41]. This is a tremendous burden [181].

3.4.2 Transputer Approach

Considering the complex tasks and real time requirements of a mobile robot as well as ease of implementation, we have therefore developed a novel computer architecture for real-time sensing and control using a new generation of processors and languages that differ from the conventional ones described above: ***transputers*** and the ***occam*** language[135].

Transputer Technology

The transputer is a single-chip microprocessor, containing all the major components of a conventional SISD (single instruction single data) machine, sequential Von Neumann computer. However in addition it has memory and a communication unit on-chip, connected through a 32-bit internal bus. The bus also connects to the external memory interface, enabling additional local memory to be used. Moreover, the floating point transputers have an on-chip floating point unit with small size and high performance [136].

There are three main important features in a transputer: (i) it has a microcoded scheduler which enables any number of concurrent processes to be executed together and share the processor time. The scheduler ensures that inactive processes (waiting for input or output, or timing) do not consume any processor time and active processes held on a list are executed one by one. Whenever a process is unable to proceed, its instruction pointer is saved in its workspace and the next process is taken from the list. (ii) communication between processes is achieved by means of chan-

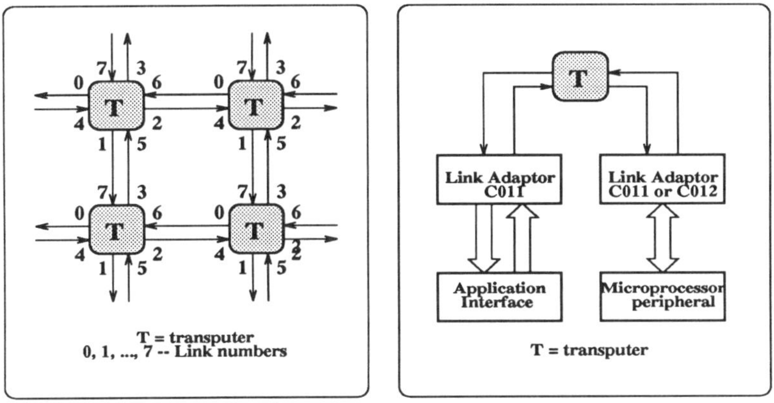

(a) (b)

Figure 3.8: A transputer network and peripheral interfacing

nels which are point-to-point, synchronised and unbuffered. A channel between two processes executing on the same transputer is implemented by a single word in memory. In contrast, a channel between two processes executing on different transputers is implemented by point-to-point, on-chip links operating at a high speed (10 or 20 Mbits/s). (iii) A multiprocessor architecture built using transputers does not need external communication facilities. Instead, it can easily be built by connecting transputers through links, as shown in Figure 3.8(a). It can also be simply interfaced to external peripheral via the specially designed Link Adaptors, shown in Figure 3.8(b).

Occam Language

Occam is well known for its influence on the design of the transputer [134]. As one of the first languages capable of automatic synchronised communication in concurrent programming [165], occam provides explicit parallelism, partitioning and scheduling for parallel processing on transputer [181]. It is based on the concepts founded by David May in EPL (Experimental Programming Language) and Tony Hoare in CSP (Communicating Sequential Processing) [93, 94], and is produced by INMOS [134]. CSP is a formalism that can be used to prove the absence of potential deadlock. The realisation of CSP in occam is so elegant that all the necessary synchronisation and control primitives for concurrency are closely coupled with the transputer instruction set. In other words, communication between different processes in occam is built into the language itself and automatically synchronised by *channels*. A channel can pass data between two occam processes that are running concurrently either on the same transputer or on different ones, which are synchronised in the sense that the input process will wait automatically for data from the output process, and vice versa [111], [41].

Based on transputers and the occam language, we have built a real-time computer architecture for our mobile robot to implement real-time sensing, planning and control tasks. We will give more details on how the complex tasks of a mobile robot are mapped onto a transputer network in the next section.

3.5 Mapping onto a Network of Transputers

The fundamental idea of our Distributed Real Time Architecture (DRTA) is to distribute the complex tasks of a mobile robot into different sensing and control modules, and then combine them into several layered control loops to form different task-achieving behaviours in a bottom-up manner. In this section, we further address how to implement this idea on a network of transputers which will satisfy the performance requirements that were described in section 3.3.

3.5.1 Modular Approach

A modular design is the natural way to build a complex control system in terms of flexibility, expandability, robustness and the ease of implementation. Following this idea, we propose a transputer-based module, called a locally intelligent control agent (LICA), as a building block for our distributed architecture [99].

Each LICA is composed of a motherboard and a few functional components: a standard TRAM (TRAnsputer Module), an interface, and a driver, as shown in Figure 3.9(a). Note that the number of TRAMs, and the size of the interface and driver, in each LICA is completely flexible, depending on which task this LICA is to perform. The board size of LICA is also flexible, according to the need of different applications.

There are three kinds of sizes of LICA board in the system we design: a standard single Eurocard, a standard double Eurocard and the commercially available PERIMOS MB8-TIO, as shown in Figure 3.12. The PERIMOS MB8-TIO is shown in Figure 3.10. Note that there are six TRAM modules on board, which are for communications, implementing RS232 and RS422 standards, for instrumentation (a flash analogue-digital converter TRAM), for computing (two T800 processor TRAMs), and, for local autonomy, an EPROM TRAM.

In general, each LICA is an expert in its own localised function and contains its own localised knowledge source. Figure 3.9(b) shows its logical function. An interface board has one or two Inmos C011 link adaptors to provide connections between the Inmos high speed link (10 or 20 Mbits/s) and an external device. The function of the interface board depends heavily upon which kind of external device is connected. We have built a range of different interface modules, for analogue-digital conversion, digital I/O, counter-timers, serial communications or any combination of

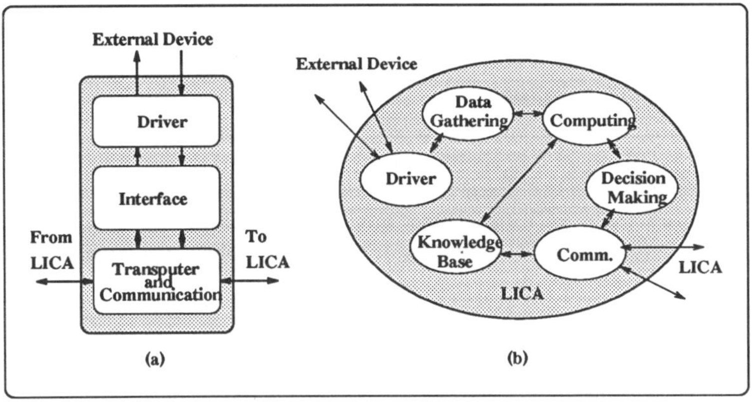

Figure 3.9: Block diagram of a Locally Intelligent Control Agent (LICA)

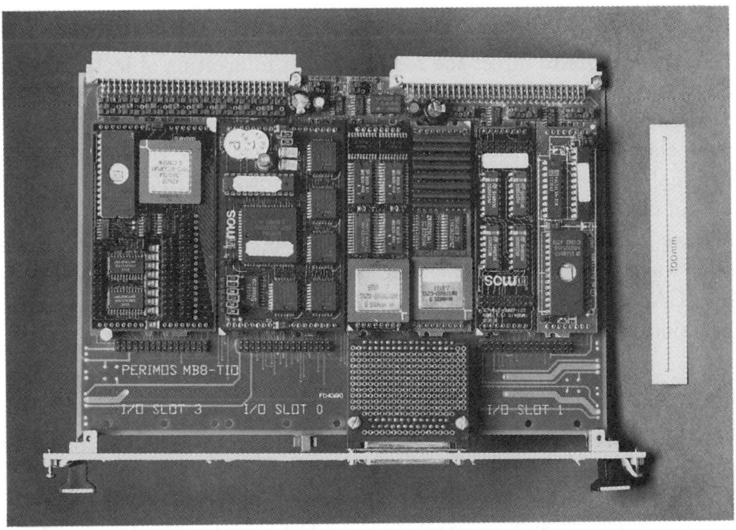

Figure 3.10: Prototype of a Locally Intelligent Control Agent (LICA)

these modules, as shown in Figure 3.12. Driver modules may be used to increase communications capability or for isolation in a noisy environment. We have designed different driver modules for our own use, such as optical isolation and RS422 modules and a motor driver.

3.5.2 Hardware Mapping

The LICA design provides an effective way to build a distributed control system for our mobile robot. Figure 3.11 shows that the complex control system of the AGV can be composed of five LICAs. Each LICA connects sensing to action to achieve a level of competence for the AGV to interact intelligently with the dynamic environment. The new controller we built includes five LICAs, as shown in Figure 3.12 (19-inch Parallax rack).

As an example, Figure 3.13(a) shows a LICA configuration for an obstacle avoidance behaviour. Here, only one transputer TRAM (size 1) has been used to fire sonars, control motors, process sonar data, build a local map, and avoid unexpected obstacles. More transputer TRAMs can be added easily if necessary. The interface board has two C011 link adaptors, connected to counters for sonar timing, digital I/O for sonar firing, and a PWM circuit for servo driving. Figure 3.13(b) shows the occam processes and program (PAR means that the processes following it are operating concurrently). **comm.proc()** procedure is in charge of communication with other LICAs. **sonar.proc()** handles sonar firing and data processing. **motor.proc()** drives the servo motors to direct the sonar towards interesting objects. **map.proc()** builds a local map to add unexpected obstacles. **avoid.proc()** works out a collision-free path when obstacles are encountered. All of them communicate with one another via software channels, $c1, ..., c4$ and $d1, d2, ..., d6$, rather than via global variables, as in conventional computer languages. Figure 3.14 shows the LICA board we built to implement an obstacle avoidance competence.

Other task-achieving behaviours such as path planning and navigator are built in a similar way. Also, more behaviours can easily be added using LICAs. In this way, the LICA design enables us to achieve flexibility, expandability, and ease of implementation.

3.5.3 Software Mapping

Occam is a block structured concurrent language, programs consisting of a network of communicating processes. The language is based on the idea of concurrency and communication. It gives us the ability to map our application onto an undetermined number of transputers, maintaining all the potential for exploiting the parallelism that exists in the application. Figure 3.15 presents the mapping of sensing and control tasks onto a network of communicating processes in the implementation of the DRTA architecture, and Figure 3.16 shows the configuration of a network of communicating processes onto 5 T800 transputers.

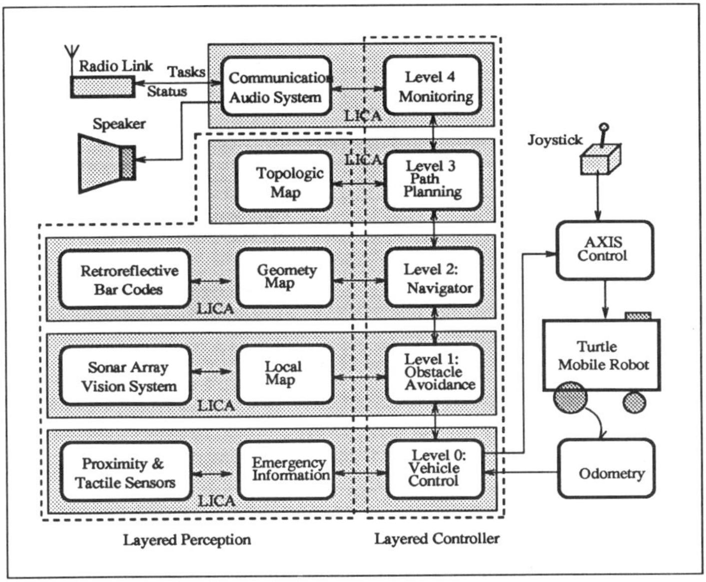

Figure 3.11: A transputer architecture for the AGV

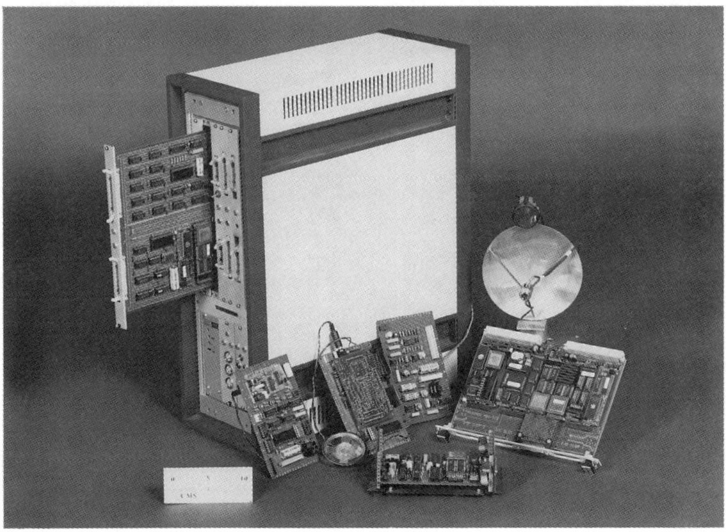

Figure 3.12: New controller consisting of five LICAs

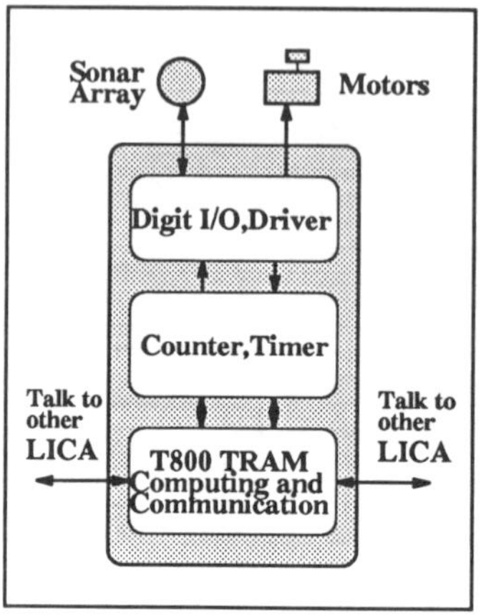

(a)

```
PROC local.planner.proc(CHAN OF ANY in1,
        out1, in2, out2, s.in, s.out, m.in, m.out)
  ... initialization
  ... sonar.sampling
  ... motor.driver
  ... map.building
  ... communication
  ... avoid.control
  PAR
    sonar.proc (s.in, s.out, d1, c1)
    motor.proc (m.in, m.out, d2, c2)
    map.proc (c1, c2, c3, c4, d3, d4)
    avoid.proc (c3, c4, d3, d4, d5, d6)
    comm.proc (in1, out1, in2, out2, d5, d6)
:
```

(b)

Figure 3.13: Block diagram and software of a LICA for a level of competence

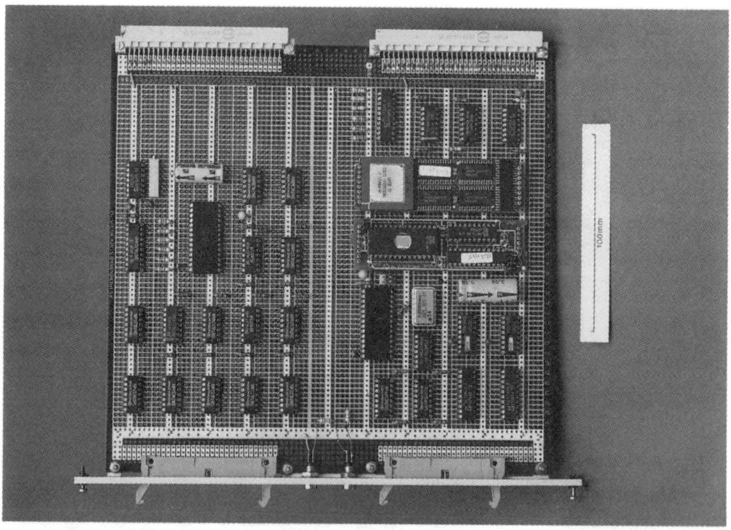

Figure 3.14: A LICA board for the obstacle avoidance competence

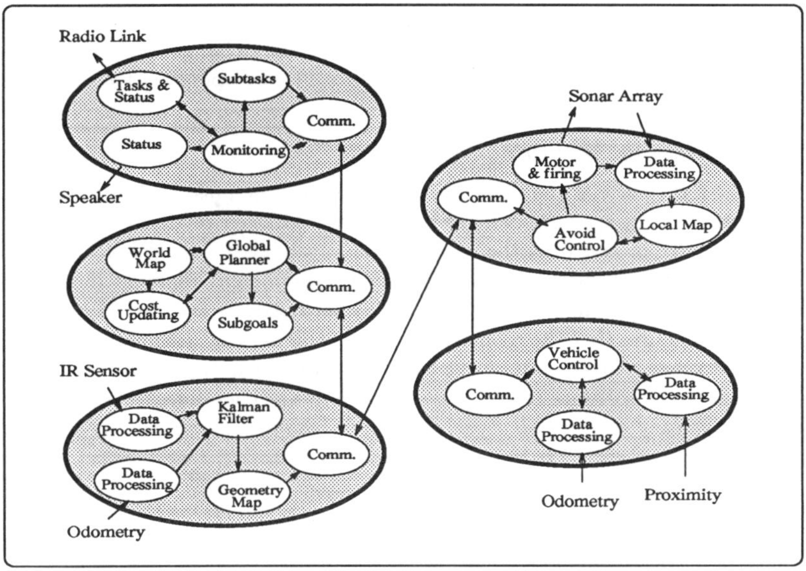

Figure 3.15: Software structure for DRTA

```
... SC monitoring.proc
... SC global.planner.proc
... SC navigator.proc
... SC local.planner.proc
... SC motion.control.proc
... initialization
PLACED PAR
  PROCESSOR 1 T8
    ... channel definition
    monitoring.proc(app1.in, app1.out, radio.in, radio.out, speak.in, speak.out)
  PROCESSOR 2 T8
    ... channel definition
    global.planner.proc(app1.in, app1.out, app2.in, app2.out)
  PROCESSOR 3 T8
    ... channel definition
    navigator.proc(app2.in, app2.out, app3.in, app3.out, ir.in, ir.out, odo.in, odo.out)
  PROCESSOR 4 T8
    ... channel definition
    local.planner.proc(app3.in, app3.out, app4.in, app4.out, sonar.in, sonar.out, mot.in, mot.out)
  PROCESSOR 5 T8
    ... channel definition
    motion.control.proc(app4.in, app4.out, odom.in, odom.out, proxi.in, proxi.out)
```

Figure 3.16: Software configuration for DRTA

To date, only five LICAs are used in the layered control architecture. As more transputers are needed, the processes currently in a transputer could be mapped easily onto separated transputers in the same LICA. The only change we have to make is that the software channels need to be mapped to the corresponding transputer links.

3.5.4 Achieving Accurate Timing

To be in the right place at the right time, the robot system must be able to determine, and compensate for, system latencies [7]. The information related by a sensor that an object is three metres from the robot is soon invalidated if the robot or the object is moving. Therefore, any sensor reading in a dynamic environment must be time stamped to define the only time at which the information is accurate. Implementing this strategy requires that the robot controller be able to measure precisely an event's time of occurrence. A transputer has a timer (a clock which 'ticks' every microsecond) that is readable by software (a *read timer* instruction). Therefore, time stamping the sensor data is easy to achieve. Synchronisation between transputers is enforced by software initially when the occam code is down loaded.

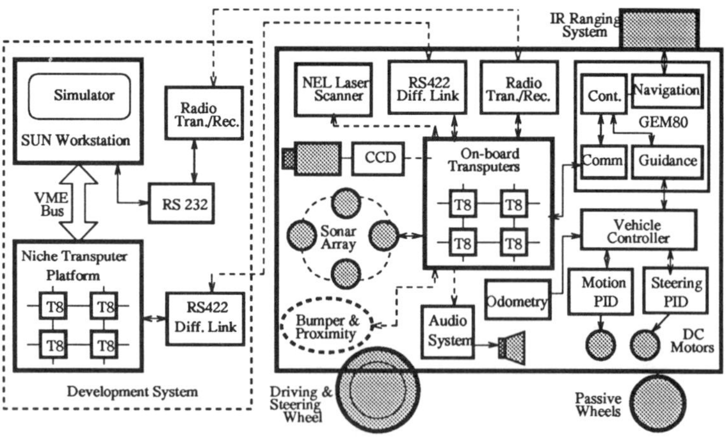

Figure 3.17: Real-time control architecture

3.5.5 Asynchronous Communication

Each LICA incorporates a task-achieving behaviour. It monitors its input data from the sensors and other LICAs, processes that data, and outputs its final decision.

Communication between LICAs are loosely coupled, using asynchronous communication. To implement this scheme, we introduce two buffer processes: a first-in-first-out buffer and a memoryless buffer for urgent communications.

3.6 Sensor-based Control Structure

Figure 3.17 shows the real-time control structure which implements sensor-based control tasks. A development environment based on the MCP 1000 (Niche) transputer platform and a SUN workstation is also included. Note that the RS422 differential links are not connected when the robot is moving. A simulator running on the SUN workstation [96] is able to control the motion of the AGV and log any data that is useful for analysis via radio links. In this way, the AGV can travel around in our AGV Lab freely without an umbilical.

Chapter 4 describes the realisation of the planning and obstacle avoidance competences implemented in this architecture.

3.7 Conclusions

The system design of mobile robots poses a key challenge to the robotics community. The system design in this chapter is heavily influenced by the need for real-time sensing, control and decision-making in order for the AGV to interact intelligently with a dynamic environment.

- The architecture we have built for the AGV supports a distributed community of sensing, planning, and action modules. Acting as a distributed problem solving network, it integrates high-level decision making with low-level control and sensor interpretation to provide task directed behaviour supported by navigation, dynamic replanning and exploration in a dynamically changing environment.

- The system is based on locally intelligent control agents (LICAs), each with its own processing unit and local memory, which do not require any common buses or central communication facility. Point to point synchronised communication links avoid contention in the communication mechanism, regardless of the number of transputers in the system. There is no global shared memory. The communication bandwidth does not saturate as the size of the system increases.

- The system has many desirable properties including modularity of sensing and control devices, flexibility for the addition of more sensing and control nodes, and high bandwidth.

Chapter 4

Sonar Directed Planning: A Decision Theoretic Approach

Huosheng Hu, Michael Brady, Penelope Probert

4.1 Introduction

In this chapter we describe methods of local planning and obstacle avoidance directed by the sonar array on the AGV. Obstacle avoidance is a key issue in achieving successful applications of mobile robots in a real world. Research on the problem can be classified as one of two approaches: *model-based* and *sensor-based.* Generally speaking, the model-based approach considers obstacle avoidance in a global way, as we decribe in later chapters of the book. It uses built-in models to describe known obstacles completely in order to generate a collision-free path and normally takes place at the global path planning level. In contrast, the sensor-based approach aims to detect and avoid unknown obstacles in real-time. This chapter addresses this issue using a sonar array to plan local paths.

Many methods for real time obstacle avoidance for mobile robots and manipulators have been reported. Among these methods, those based on artificial potential fields are the most popular, such as the *Potential Field Method* [117], the *Generalized Potential Field* [122], the *Force Field* [33], [9], the *Virtual Force Field* [25], and the *Goal Relocation Approach* [1].

The potential field method is an attractive approach to the collision avoidance problem because of its simplicity and efficient implementation. However, the complexity of tasks that can be implemented using this approach is limited. In a cluttered environment, the robot may become trapped because of the occurrence of local minima in the resultant potential field. This occurs when the attractive forces of the goal are balanced by the repulsive forces due to the presence of obstacles, a problem particularly likely to occur in environments including concave obstacles. Other problems include path oscillations when travelling through a narrow passage or corridor, and the inability to pass easily among densely spaced obstacles [120]. To overcome these problems, Borenstein and Koren [27] proposed another obstacle avoidance method called the *Vector field histogram*, which consists of a two-dimensional Cartesian *Histogram grid* (at world-model level), updated continuously with range data, and a one-dimensional *Polar histogram* (at intermediate data-level) that represents the polar obstacle density around the robot. A sector with low obstacle density which is close to the goal direction is selected as the robot's direction of motion. Zhao and BeMent [199] developed a heuristic search-based algorithm for trap recovery.

However, most of the potential field methods mentioned above are in fact difficult to implement directly on mobile robots whose kinematic constraints are nonholonomic, such as the AGV (but see Chapter 11). The main reason is that these mobile robots have a lower bound on the turning radius and are unable to follow an arbitrary path or change direction instantaneously. Moreover, limitations on robot motion within roadways in our case imposes new problems which are not taken into account by potential field methods.

In the next section we present a framework for factory planning, using a simple global model. In formulating plans we aim to make the best decisions under the uncertainty of sensor readings and we present a decision theoretic approach to solve the obstacle avoidance problem. We describe how heuristic interpretation rules and probabilistic sensor models are built to reduce the uncertainty in sonar data. Succeeding sections place an obstacle avoidance algorithm in the framework of a loss function and the recursive Bayes decision rule. Finally we present results on implementation.

4.2 Framework for Planning

In a typical factory a vehicle's route is represented by roadways and connecting junctions. We adopt this representation for our global planner, using a static, topological map of the factory layout. However the use of dynamic cost factors on the nodes of the map allows the robot to take into account sensory data in choosing the best route.

4.2.1 Environment and its Topological Map

First we describe the environment in which the robot operates and the global planner. Our test site is 15m long and 3.8m wide, and includes four pillars. We assume that the geometrical features of this environment are known *a priori*, but people, tables, chairs, boxes, *etc.* form unknown obstacles which may be in any position. Therefore, we have a partially-known environment where the mobile robot has to cope with unexpected obstacles during its missions.

Real-time path planning requires a simple representation and we construct two world maps: one to represent the global features of the environment and one the immediate region around the robot to plan a collision-free trajectory. Separation of map building into two levels is a natural and effective way to realize the two level path planning strategy we adopted.

We construct a topological map that contains information about how the different parts of the world are inter-connected. This information is typically invariant for the environment and allows the robot to plan a feasible path and verify its location as it travels. The map is a simple non-directed graph consisting of a network of places (nodes) classified as rooms, corridors, corners and junctions. Figure 4.1 shows the topological map of our AGV Lab generated from its geometrical description. It is formulated by taking into account the robot's size, the topological and geometrical features of the environment, and the ranges of the sensors used. Note that since rooms and corridors are unlikely to change, the topological map can be regarded as fixed.

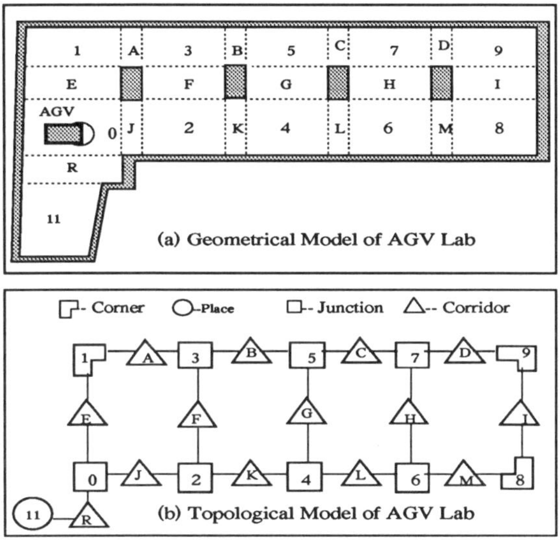

Figure 4.1: Geometrical map and topological map

The map is updated, and local planning implemented, using information from a sonar array.

4.3 Sensor Configuration

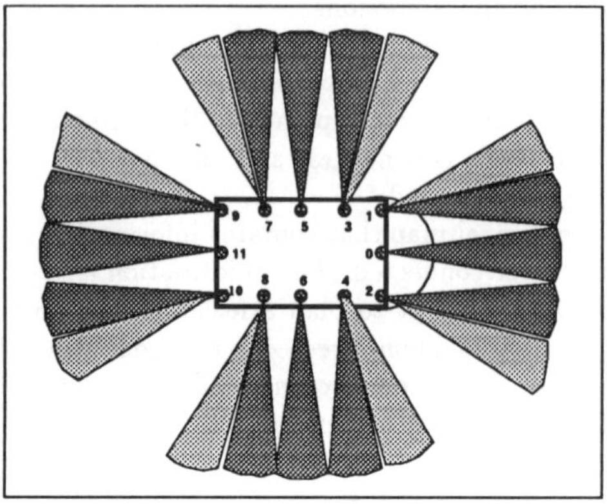

Figure 4.2: The configuration of an array of 12 sonars

To achieve real-time adaptivity, we have built a "sonar curtain" consisting of an array of 12 Polaroid sonars around the robot, shown in Figure 4.2. Each of them is driven by a Futaba servo motor and can be rotated up to 170 degrees. The rotation speed is about 100 degrees per second. The light shaded cone depicts one sonar beam rotated about one beam width from the original direction shown as the heavily shaded cone. This provides greater flexibility than the more usual fixed sonar array such as [146], [50], [33], [191], [56], [127], [27]. To avoid crosstalk (the sound wave from one transducer influencing others), the 12 sonars can not simply be operated simultaneously. Instead, we adopt a group sampling strategy. The 12 sonars are spatially separated by 30cm, and also decoupled in time in the sense that only one group of four sonars, one per side, is sampled at the same time. The other two groups are waiting or rotating. The maximum range of sonar sensors is set to be $R_{s\max} = 4.2m$. Any return range that is more than $R_{s\max}$ will be ignored so that the multiple reflection problem can be effectively reduced. To read all 12 sonar sensors in three groups takes about 120ms.

4.3.1 Local Map and Collision Zone

We introduce a local map to record the sensory information provided by the 12 sonar sensors, based on the robot local frame. The map is composed of a collision zone and a prediction zone, as shown in Figure 4.3. The collision zone is a two-dimensional circular area with radius R_0 (proportional to the robot speed), located at the physical center of the vehicle. It is divided into four subzones: front, back, left, and right. When the sonar sensors see an unexpected obstacle in one of the subzones along which the robot is currently moving, a control action such as *stop* or *shyaway* will be chosen immediately to avoid collision. In contrast, the prediction zone is a rectangular space (measuring $4m \times 2.7m$) immediately ahead of the robot, in the direction which the robot is approaching. It is used mainly to predict potential collisions based on noisy sensor data, and to choose control actions such as a_1 (manoeuvre around an obstacle) or a_2 (backtrack).

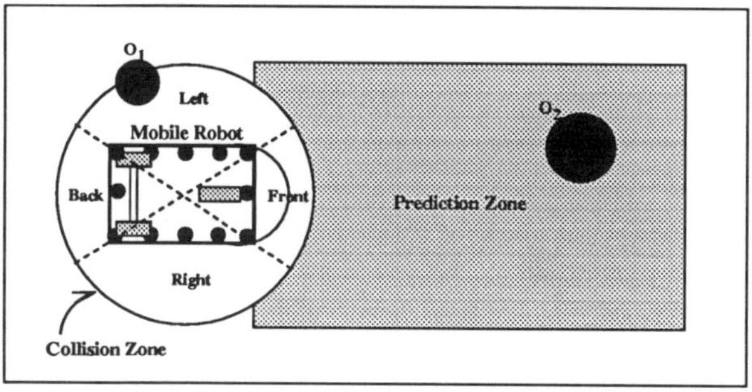

Figure 4.3: The local map consisting of collision and prediction zone

Since sonar sensor readings typically cover a large space, it is necessary to provide a mapping from the real sensor space to the abstract sensor space. In other words, the real sensor data is abstracted into: (i) short range, which falls into the collision zone and prediction zone and means that there is an obstacle close by, and (ii) long range, which means there probably is no obstacle. In this way, the robot is able to react to unexpected obstacles very quickly. The abstracted short range data is mapped into a local map and further processed for local obstacle avoidance in real time. For instance, the obstacle O_1 in Figure 4.3 is mapped into the left subzone, but no action is taken since the robot is moving forward. Instead, the robot reacts to the obstacle O_2 in the prediction zone and chooses a suitable action to avoid it.

4.3.2 Rule-based Interpretation of Sonar Data

The sonar are used to direct a local obstacle avoidance algorithm and to update the global map. Both applications require them to distinguish between three situations: (i) a clear pathway, (ii) the presence of an obstacle about which a sidestep maneouvre can be performed and (iii) the presence of an obstacle entirely blocking the path which requires the robot to backtrack and seek an alternative route.

Therefore, confronted by an unexpected obstacle in the prediction zone, the robot has to use its three front sonars to estimate the width of the gap, to judge whether it is passable. Since the wide beam width of sonar sensors make accurate measurements of the clearance of the path difficult, we adopt a simple heuristic rule for interpreting the three sonar readings. It is based on the superimposition of their beam wavefronts at some distance, as shown in Figure 4.4, where the robot is moving into a portion of path constrained by the road boundary.

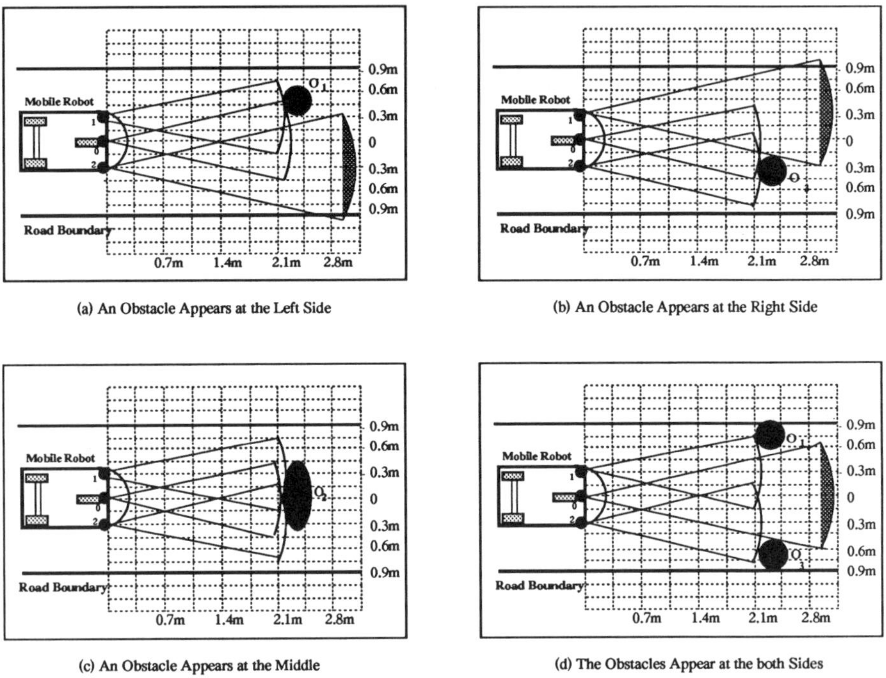

Figure 4.4: Interpretation of three front sonar data. Note that the light shaded circles meant that nothing is found in this part of prediction zone.

The three beams superimpose at ranges greater than $0.7m$, as shown in Figure 4.4. We choose the range $2.1m$ as the decision distance for the robot, chosen from the robot's speed $0.5m/s$, the road width $W_r = 1.8m$, and the resolution that can be provided by the three front sonars.

The heuristic rules for interpretation can be simply described as:

Rule 1 If all of three sonars see clearance, the predetermined path is passable and no action should be taken.

Rule 2 If the middle sonar sees nothing in the cone of its acoustic beam, then the predetermined path is passable even though the two side sonars may see obstacles, as shown Figures 4.4(d).

Rule 3 If any one of both side sonars sees clearance in the prediction zone and the road width W_r is at least twice the robot width, the gap is considered wide enough for a manoeuvre although the other two sonars may see an obstacle, as shown in Figures 4.4(a) and 4.4(b).

Rule 4 If all three sonars see something, then the path is announced to be a blockage, as shown in Figure 4.4(c).

This rule-based interpretation of the three sonar data is obviously very simple. However, there are two problems that we have to solve so that it can be used in practice. The first problem is that the feasible path may be eliminated by the broad beam width of the sonars. For instance, Figure 4.5(a) shows that the right side gap is actually passable even though three sonars see the obstacle O_2. We name this situation as a 'miss' measurement since rule 4 concludes that the path is impassable. The second problem is caused by specular reflection. Figure 4.5(b) presents a situation in which the obstacle O_1 can not be detected by the sonars since its surface is not perpendicular to the acoustic axis of the sonar beam. As a result, an impassable path is interpreted as passable. We call this situation a 'false' measurement.

To solve the 'miss' and 'false' interpretations, it is necessary to use a probabilistic measure for the rules to represent knowledge about sensor errors. The probability measure is performed by 5000 random trials. The main obstacles we consider are furniture, boxes, and people. We assume that the size of the obstacles (as seen from the sensors), O_s, is a random variable with uniform distribution between 10cm and 60cm. The main features of the obstacles include the following: corners, edges, cylinders, planes, and coarse irregular surfaces. The appearance of these features is also assumed to be a random variable that is uniformly distributed. We introduce the following terminology:

- N_t is the total number of trials (measurements).
- $n_1 \in N_t$ is the number of the times the state is θ_1 (passable).
- $n_2 \in N_t$ is the number of the times the state is θ_2 (impassable).
- $\mathbf{z}_{1c}$ is the correct measurement given that the state is θ_1, and n_{1c} is the number of times the event $\mathbf{z}_{1c} \in n_1$ occurs.

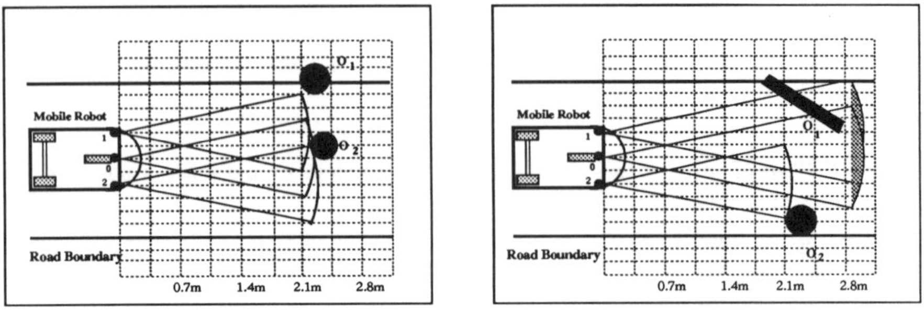

(a) Three Sonars see an Obstacle

(b) The Obstacle 1 is not Detected

Figure 4.5: Problems caused by the wide beam width and specular reflection

- $\mathbf{z}_{1m}$ is the miss measurement given that the state is θ_1, and n_{1m} is the number of times the event $\mathbf{z}_{1m} \in n_1$ occurs.
- $\mathbf{z}_{2c}$ is the correct measurement given that the state is θ_2, and n_{2c} is the number of times the event $\mathbf{z}_{2c} \in n_2$ occurs.
- $\mathbf{z}_{2f}$ is the false measurement given that the state is θ_2, and n_{2f} is the number of times the event $\mathbf{z}_{2f} \in n_2$ occurs.

From the above definitions, we have

$$N_t = \sum_{i=1}^{2} n_i = n_{1c} + n_{1m} + n_{2c} + n_{2f} \tag{4.1}$$

and the conditional probabilities of the four events $\mathbf{z}_{1c}$, $\mathbf{z}_{1m}$, $\mathbf{z}_{2c}$, and $\mathbf{z}_{2f}$ are defined as

$$\begin{array}{ll} P(\mathbf{z}_{1c}|\theta_1) = n_{1c}/n_1 & (\text{for the event } \mathbf{z}_{1c}) \\ P(\mathbf{z}_{1m}|\theta_1) = n_{1m}/n_1 & (\text{for the event } \mathbf{z}_{1m}) \\ P(\mathbf{z}_{2c}|\theta_2) = n_{2c}/n_2 & (\text{for the event } \mathbf{z}_{2c}) \\ P(\mathbf{z}_{2f}|\theta_2) = n_{2f}/n_2 & (\text{for the event } \mathbf{z}_{2f}) \end{array} \tag{4.2}$$

It is obviously that the following conditions hold:

$$P(\mathbf{z}_{1c}|\theta_1) = 1 - P(\mathbf{z}_{1m}|\theta_1) \tag{4.3}$$

$$P(\mathbf{z}_{2c}|\theta_2) = 1 - P(\mathbf{z}_{2f}|\theta_2) \tag{4.4}$$

Table 4.1 presents the results from such a trial. The probability of miss measurements is reduced as the robot approaches the obstacle closely. In contrast, the probability of a false measurement increases when the obstacle is close to the robot. This is because there is no reflection from the surface itself as it is specular. However

Table 4.1: Conditional Probability of Sonar Reading Given the State θ

Range (meter)	Correct $P(\mathbf{z}_{1c}\|\theta_1)$	Miss $P(\mathbf{z}_{1m}\|\theta_1)$	Correct $P(\mathbf{z}_{2c}\|\theta_2)$	False $P(\mathbf{z}_{2f}\|\theta_2)$
1.40	0.82	0.18	0.72	0.28
1.75	0.73	0.27	0.75	0.25
2.10	0.65	0.35	0.79	0.21
2.45	0.57	0.43	0.83	0.17
2.80	0.49	0.51	0.86	0.14
3.15	0.41	0.59	0.89	0.11
3.50	0.33	0.67	0.91	0.08

at a distance the sonars see a return from the edges, whereas at close range the edges are outside the beam width and the sonars see nothing. From this point of view, the sonar wide beam width is good for obstacle detection at some distance.

The sonar data is used for two purposes: updating the global model for the global path planner and in local obstacle avoidance.

4.4 Global Path Planner

The topological world map we construct is designed primarily for fast, convenient storage and abstraction of geometrical information about the real world. It is used for the global path planner to determine a route or a set of intermediate points for the robot to follow from a start position located in one place to a destination located in another place without collision. We use minimum time (rather than distance) as a metric to determine an optimal path among the alternatives. This is because a minimum distance path may cause the robot to decelerate sharply at corners or stop to pivot, and may also not be safe as it passes too near to obstacles.

We use the roadways to represent the free motion space for our mobile robot. The road site consists of walls and pillars at a safe distance. Figure 4.6 shows the central lines of the roads defined in the topological map (figure 4.1) but takes kinematic constraints into account. As we can see, these predefined roads consist of straight lines and arcs which in turn form the edges of the search graph we describe in the following section.

4.4.1 Cost Function

Global path planning starts with the selection of the cost function to be used to assign merit to a path. To take into account the probability of encountering unexpected obstacles during path planning, we define the cost function to consist of two parts: one is deterministic cost (time spent without obstacles), the other involves uncertainty (time spent on avoiding obstacles). The deterministic cost depends on the path

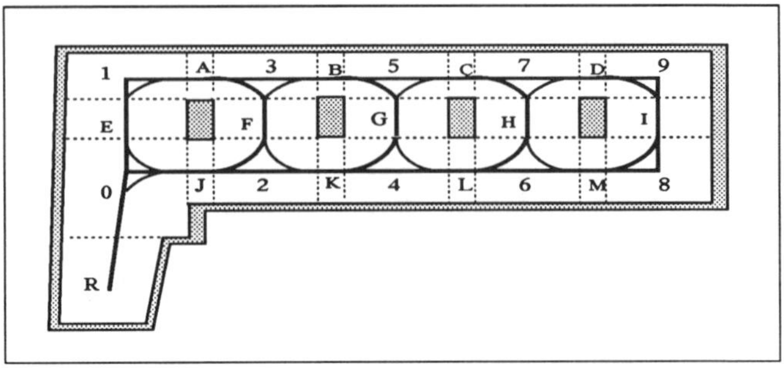

Figure 4.6: A road representation

parameters such as length, width, and straightness. Uncertainty cost is estimated by assuming a prior distribution for unknown parameters and varies dynamically according to information accumulated by sensors. In this way, the mobile robot learns global knowledge from its travels.

Suppose that the mobile robot moves at a piecewise constant speed, corresponding to different path situations such as a straight line or arc. Let C_{ij} be the cost associated with an arc or a straight line from node i to node j, defined as follows:

$$C_{ij} = t_{ij} + \sum_{n} u_{ij}(n) \tag{4.5}$$

where: t_{ij} — the time spent when there are no obstacles between nodes i and j.
u_{ij} — the uncertainty cost between nodes i and j; 0 for no obstacle, otherwise > 0.
n — the number of unexpected obstacles.

Based on the cost function defined above, a non-directed search graph with cost weights on edges is shown in Figure 4.7. These cost weights can be initially estimated based on prior information and are then updated on the basis of sensor data. Dynamic programming is used to find the route with the minimum cost function.

Given the planning framework, we need only record how many and which kind of obstacles the robot has met during its traversal to update the uncertainty cost of each portion of path for the use of global planning.

4.4.2 Updating the Global Model

Now consider how to update the map from sensor data as the robot moves around. We update the map through altering the cost functions associated with each path segment in the weighted search graph of figure 4.7.

We define three types of obstacle: O_1 around which it is possible to make a local maneouvre within the same pathway, O_2 which blocks a pathway completely so a backtrack is required and O_3, a moving obstacle. In this chapter we consider only

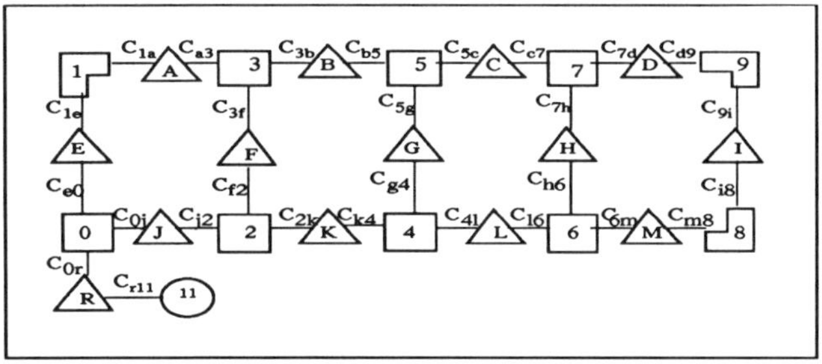

Figure 4.7: a weighted search graph

the first two types of obstacle; moving obstacles are discussed in [98]. We use a probabilistic approach based on Bayes statistics to update the current hypothesis on obstacles and hence the cost factors associated with the relevant parts of the route.

Consider first the occurrence of an obstacle of type O_1, round which maneouvres are possible. We allow several obstacles O_1 to be present on a segment of the path, (i, j). We choose a *Poisson* distribution to model the number of instances, W_1, of O_1 we might expect in the path segment:

$$p(W_1 = k|\lambda) = \frac{\lambda^k}{k!}e^{-\lambda} \tag{4.6}$$

where $\lambda > 0$ is initially unknown.

We assume that λ has a conjugate prior density function of the *Gamma* distribution. We use the *Gamma* distribution since it maps easily to our application with the interpretation of its parameters as (i) the number of obstacles encountered on each portion of the path and (ii) the number of traversals of the path.

Similarly we model the presence of an obstacle O_2 by a discrete random variable W_2. Only one obstacle need be present to block the path completely and so we use the *Bernoulli* distribution for W_2 to model the presence or absence of an obstacle p is the probability of encountering an obstacle along the path:

$$f(W_2 = k|p) = p^k(1-p)^{1-k} \quad (0 \leq p \leq 1) \tag{4.7}$$

where k is zero if no obstacle is present and one if there is an obstacle.

We assume that the conjugate prior density of p follows the *Beta* distribution and agin identify its parameters with observations: (i) the number of successful traversals between i, j and (ii) the number of unsuccessful traversals.

The parameters describing these distributions are updated using Bayes statistics as the robot moves around and new information becomes available from the sonar array. We can therefore determine dynamically the expectation of encountering each type of obstacle: $E(O_1)$ and $E(O_2)$. We then update the uncertainty cost of any

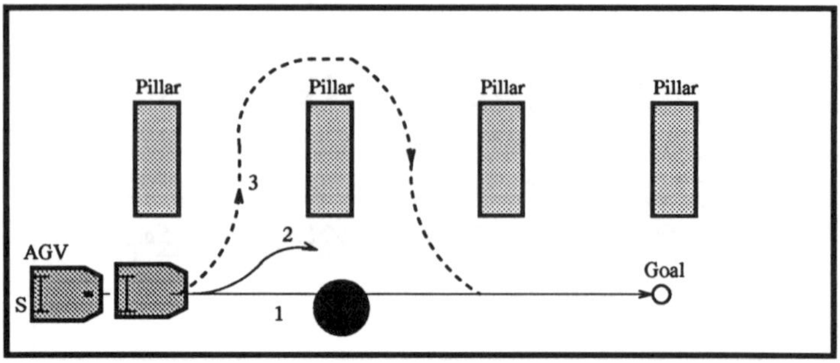

Figure 4.8: The AGV encounters an unexpected obstacle

section of the weighted search graph i, j according to the expectation of encontering each type of obstacle and the cost associated with each c_1 and c_2:

$$u_1 = c_1 E(O_1) \qquad u_2 = c_2 E(O_2) \tag{4.8}$$

The c_i are chosen heuristically through considering factors such as path length, time of traversal and safety. A decay factor is built into the uncertainty cost so that if the robot makes no observation on part of the route for some time the cost associated with that part eventually return to its default value.

Further details and results on the application of this method are given in [97].

4.5 Changing Environment: Problem Fomulation

Now consider more generally how to plan in a changing environment. Assume that the mobile robot is travelling along an optimal path generated by the global path planning algorithm shown as path 1 in Figure 4.8. However, an unexpected obstacle appears along its path and there arises the possibility of a collision. As we saw earlier, the sensors can only estimate the probability of the gap between the obstacle and the pillar being wide enough for a sidestep manoeuvre. Therefore, a decision problem arises: should the robot continue its motion along path 2 to make further observations with the expectation of performing a sidestep manoeuvre or simply backtrack along path 3? Both decisions incur a cost. If the cost for extra sensing is less than the cost for taking the alternative path, it may be worth persevering along path 2. But the robot may eventually find the gap is impassable, incurring an overall cost greater than following path 3.

To solve the problem described above, we face two basic difficulties: (i) how to deal with uncertainty in collision detection, and (ii) how to choose an optimal action based on imperfect state information. The first is met through the probabilistic model for the sonar array. To handle the second we seek a decision rule for an *avoiding* control task in the face of uncertainty.

We consider the appearance of the unexpected static obstacles to be a random event. Let $\mathbf{\Theta}$ denote the state of the path to be a random variable consisting of two states:

$$\mathbf{\Theta} = (\theta_1, \theta_2) = (passable, impassable) \tag{4.9}$$

Given that the mobile robot should stay on the path in any case, strategies to avoid unexpected static obstacles include: (i) *manoeuvre*, if there is a large enough gap, (ii) *shyaway*, when an obstacle is too close, (iii) *backtrack*, if necessary, and (iv) *stop*, in an emergency. Note that both *shyaway* and *stop* are for emergency use only. Therefore, we define the action space by:

$$\mathcal{A} = (a_1, a_2) = (manoeuvre, backtrack) \tag{4.10}$$

The robot chooses an action a from a set $\mathcal{A}$ of all possible actions based on the observation $\mathbf{z}$ of the current state of the path θ. This is a decision process mapping the sample space onto the action space $\mathcal{A}$, denoted by $a = d(\mathbf{z})$. In general, the decision made by the robot is concerned with the best possible action in the face of uncertain information $\mathbf{z}$. Therefore, a way of evaluating a decision or action is required. We define a *loss function* $L(\theta, a)$ which gives a measure of the loss incurred in taking the action a when the state is θ. It is often chosen subjectively in order to reflect the merit of the chosen action for the given θ. For example, when the state $\theta = \theta_2$ (impassable) and the robot chooses backtracking action a_2, then the incurred loss $L(\theta_2, a_2)$ can be defined as zero. The most desirable action for the robot is that which minimizes $L(\theta, a)$ when the state of nature is θ. We assume throughout this chapter that the loss function is bounded below; then, without loss of generality, we assume that the loss function we use is nonnegative

$$L(\theta, a) \geq 0 \tag{4.11}$$

The way the action is chosen by the robot is called the decision rule. A Bayes decision rule is pursued here to minimise the expected loss over the possible actions, based on estimates on the environment from the sensors. The sensors take measurements $\mathbf{z}$ of the value of the state θ. Following a sensor reading, we use the posterior estimate of θ in our decision making:

$$p(\theta|\mathbf{z}) = \frac{p(\mathbf{z}|\theta)p(\theta)}{\sum p(\mathbf{z}|\theta)p(\theta)} \tag{4.12}$$

where $p(\mathbf{z}|\theta)$ is dicussed in section 4.3 and $p(\theta)$ is the prior information on the state [42]. We then calculate the *Bayes risk* for each action a being chosen[15]:

$$B(p(\theta|\mathbf{z}), a) = E^{p(\theta|\mathbf{z})}[L(\theta), a] = \sum_{\mathbf{\Theta}} L(\theta, a)p(\theta|\mathbf{z}) \tag{4.13}$$

It is obvious that the Bayes risk is the average of the expected loss (also called the *risk function*) with respect to the prior distribution of θ for each action a. Note that it depends on the true state of nature, θ, and the action, a. Then, the Bayes decision rule is the selection of an action a that minimizes the Bayes risk. More precisely, the Bayes rule chosen by the robot is the action $\hat{a}$ which minimizes the posterior Bayes risk:

$$\hat{a} = arg \min_{\mathcal{A}} B(p(\theta|\mathbf{z}), a) \tag{4.14}$$

Whenever sensor data is available we use a decision-making algorithm for finding the Bayes rule as follows:

Step 1 Wait for the current observation $\mathbf{z}$ to be provided by the heuristic rules for interpreting sonar data.

Step 2 Calculate the posterior distribution, $p(\theta|\mathbf{z})$, of the state.

Step 3 For each action a, calculate the Bayes risk $\sum_{\Theta} L(\theta, a)p(\theta|\mathbf{z})$.

Step 4 Find the Bayes rule: that is the action $\hat{a}$ with a minimum Bayes risk.

Step 5 Return to **Step 1** and continue until the robot reaches the goal.

4.5.1 Formulation of Loss Function

Determination of the Bayes rule requires values for the loss functions $L(\theta, a)$ over the action space. The function $L(\theta, a)$ is a real value which is assigned as the outcome of action a when the state is θ. In our case, both the state space Θ and the action space $\mathcal{A}$ have two components. Therefore, we have four loss functions to be defined: $L(\theta_1, a_1)$, $L(\theta_1, a_2)$, $L(\theta_2, a_1)$, and $L(\theta_2, a_2)$.

$L(\theta_1, a_1)$ is the loss incurred by taking the action a_1 (manoeuvre) given that the path is passable, $\theta = \theta_1$. We simply define it to be zero since the robot lost nothing. Similarly, we have $L(\theta_2, a_2) = 0$, which means that the robot does not suffer any loss when it chooses to backtrack given that the state is impassable, $\theta = \theta_2$. (Recall that the robot does not know the true state θ when it makes its decision. Instead, the decision made by the robot is based on a probabilistic interpretation of the sonar data.)

Both $L(\theta_1, a_2)$ and $L(\theta_2, a_1)$ are not equal to zero. $L(\theta_1, a_2)$ is defined in terms of the loss incurred by *backtrack* when the state of the path is in fact passable. It can be calculated by

$$L(\theta_1, a_2) = C_a - C_c \tag{4.15}$$

where C_a is the cost of travelling along the alternative path from the current position to the goal, and C_c is the cost of travelling along the predefined path. Both of them are sent from the global planner dynamically.

The loss $L(\theta_2, a_1)$ is incurred by the *manoeuvre* action when the true state of the path is impassable, and is defined by

$$L(\theta_2, a_1) = C_o + C_m \tag{4.16}$$

Action	θ_1	θ_2
a_1	0	$C_o + C_m$
a_2	$C_a - C_c$	0

Table 4.2: Definition of the Loss Function

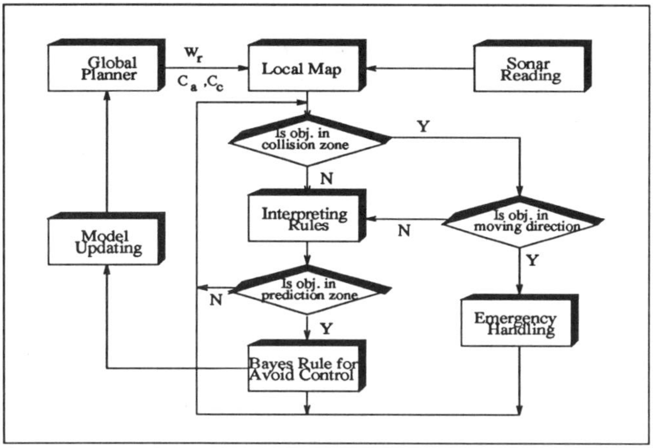

Figure 4.9: Flow chart of implementation of obstacle avoidance algorithm. Note that W_r is the road width of the current portion of path. C_a and C_c are the costs of travelling along the alternative path and the predefined one respectively

where C_o is the cost needed to take the extra observation, and C_m is the cost incurred by travelling from the decision position to the position where the extra sensing is taken.

The loss function is therefore represented by a *loss matrix* as shown in table 4.2.

4.5.2 Obstacle Avoidance Implementation

Figure 4.9 is a flow chart of the obstacle avoidance algorithm. We can clearly see how the decision rule finding algorithm is incorporated. As soon as the three front sonars see an unexpected obstacle in the prediction zone, the Bayes rule is selected to minimise the expected loss so that the optimal control action can be taken to avoid collisions. The implementation results are sent back to the global path planner to update the uncertainty cost which we have defined in [97].

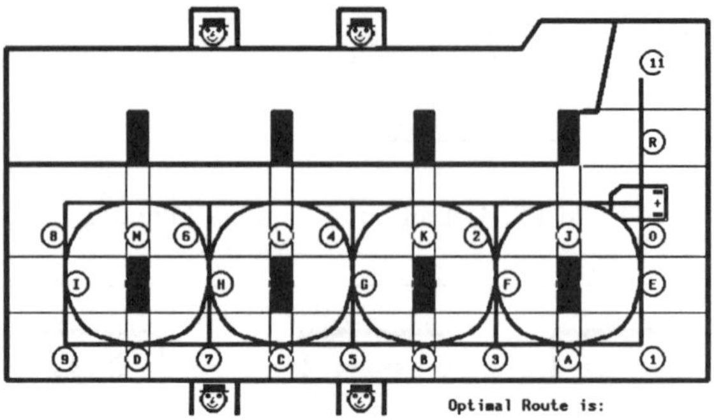

Figure 4.10: The model and partition of Oxford AGV Laboratory

4.5.3 Examples of Implementation

The proposed algorithm has been implemented both by simulation on a SUN workstation and in reality using the distributed real time architecture (see Chapter 3) of the AGV. Figure 4.10 shows the model of the AGV Laboratory, the partition of its free space, and the central line of the defined roadway. Here, we give a couple of examples.

4.5.4 Example 1

The mobile robot is commanded to move from the start position, *node 0*, to the goal, *node 8*. The optimal path planned by the global path planner is a straight line from *node 0*, via the intermediate nodes: *node J*, *node 2*, *node K*, *node 4*, *node L*, *node 6*, and *node M*, to *node 8*, which has a cost $C_c = 20s$ in terms of the path length $12m$ and the robot's speed $0.6m/s$.

When the robot moves along this optimal path, an unexpected obstacle appears on the way, see Figure 4.11. The heuristic rule 3 concludes that the state of the path is passable, i.e. $\theta = \theta_1$, since the left sonar (sonar 1) sees clearance. A new subgoal located at the middle of the gap near the obstacle is given. Which action should the robot take: *manoeuvre* or *backtrack*? We show how the robot finds the Bayes rule to give the answer, based on the proposed strategy.

Recall that the interpretation is made at a range $2.1m$ of sonar readings. From Table 4.1, the confidence we have on the current observation $\theta = \theta_1$ is:

$$p(\mathbf{z}|\theta_1) = 0.65, \quad p(\mathbf{z}|\theta_2) = 0.21$$

where $p(\mathbf{z}|\theta_1)$ is the measure of how "likely" a state of nature is θ_1 given the observation $\mathbf{z}$, and $p(\mathbf{z}|\theta_2)$ is a measure of how "likely" the state is θ_2 given the observation $\mathbf{z}$.

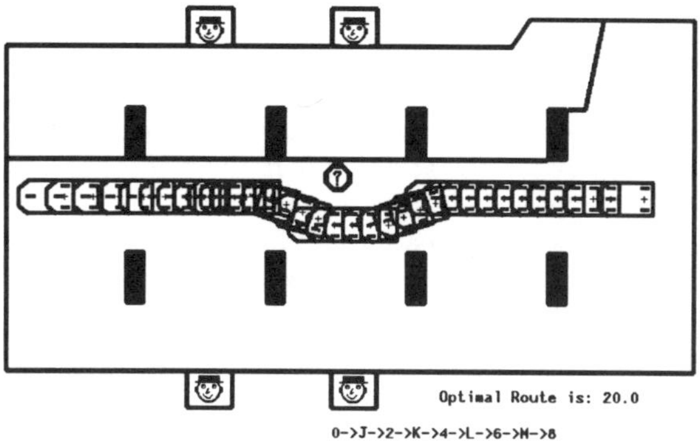

Figure 4.11: An action *manoeuvre* is chosen to minimize the expected loss

In our application, the best backtrack path has a cost $C_a = 27.2s$ (generated by the global path planner). we can calculate the loss $L(\theta_1, a_2) = 13.2s$ using equation 4.15. The loss $L(\theta_2, a_1)$ is incurred by doing extra sensing and we assume that it has the constant value: $L(\theta_2, a_1) = 8s$ (this is actually much larger than we would incur in practice for sensing with a sonar array).

Suppose that we have a priori distribution of the state θ for each portion of path:

$$p(\theta_1) = 0.7, \quad p(\theta_2) = 0.3$$

Now we first calculate the posterior distribution using Equation 4.12 to obtain the following results:

$$p(\theta_1|\mathbf{z}) = 0.88, \quad p(\theta_2|\mathbf{z}) = 0.12$$

We next calculate the posterior Bayes risk for a_1 and a_2:

$$\begin{aligned} B(p(\theta|\mathbf{z}), a_1) &= L(\theta_1, a_1)p(\theta_1|\mathbf{z}) + L(\theta_2, a_1)p(\theta_2|\mathbf{z}) = 0.96 \\ B(p(\theta|\mathbf{z}), a_2) &= L(\theta_1, a_2)p(\theta_1|\mathbf{z}) + L(\theta_2, a_2)p(\theta_2|\mathbf{z}) = 11.62 \end{aligned}$$

Comparing the two, we see that a_1 has the smaller expected loss. Therefore, the robot chooses the *manoeuvre* action to sidestep the obstacle, shown in Figure 4.11. From the above derivation, we can easily see that the decision made by the robot is based on three factors: (1) the prior information about the state is heavily weighted on $p(\theta_1)$; (2) the information provided by the sensors shows that the state θ is likely to be passable since $p(\mathbf{z}|\theta_1) > p(\mathbf{z}|\theta_2)$; (3) the loss incurred by taking the action a_2 is larger than the loss incurred by taking action a_1.

4.5.5 Example 2

The mobile robot again moves from the start position to the goal along the straight line path which has a total cost $C_c = 21.4s$ generated by the global path planner. This cost is slightly larger than in the last example because of the increase in the uncertainty cost caused by the obstacle that partially blocked the path.

Since there is an unexpected obstacle that appears in the middle of the path, the sonars see it when the robot is approaching. When the obstacle is located at a range of $2.1m$, the heuristic interpretation of the sonar data indicates that the path is impassable, $\theta = \theta_2$. Should the robot backtrack along the alternative path? We now see the answer provided by the Bayes rule. From Table 4.1, we have the confidence in this observation as follows:

$$p(\mathbf{z}|\theta_1) = 0.35, \quad p(\mathbf{z}|\theta_2) = 0.79$$

where $p(\mathbf{z}|\theta_1)$ is the probability that the current measurement $\mathbf{z}$ is a *miss* measurement, and $p(\mathbf{z}|\theta_2)$ is the probability that the current measurement $\mathbf{z}$ is a correct measurement.

The cost of the alternative path from the current position to the goal is the same as in the last example, $C_a = 27.2$. Therefore, we can calculate the loss $L(\theta_1, a_2) = C_a - C_c = 11.8s$ using equation 4.15. Following from example 1 the prior information of the state θ for current decision-making is

$$p(\theta_1) = 0.88, \quad p(\theta_2) = 0.12$$

Using Equation 4.12, the posterior distribution of the state θ becomes

$$p(\theta_1|\mathbf{z}) = 0.76, \quad p(\theta_2|\mathbf{z}) = 0.24$$

We next calculate the posterior Bayes risk for a_1 and a_2:

$$\begin{aligned} B(p(\theta|\mathbf{z}), a_1) &= L(\theta_1, a_1)p(\theta_1|\mathbf{z}) + L(\theta_2, a_1)p(\theta_2|\mathbf{z}) = 1.92 \\ B(p(\theta|\mathbf{z}), a_2) &= L(\theta_1, a_2)p(\theta_1|\mathbf{z}) + L(\theta_2, a_2)p(\theta_2|\mathbf{z}) = 8.97 \end{aligned}$$

Comparing the two, we see that a_1 has the smaller expected loss and is thus the Bayes rule. Therefore, the robot chooses the *manoeuvre* action. When the robot takes extra observations at close range, it eventually finds that the gap is impassable. The robot has to backtrack along the alternative path, as shown in Figure 4.12. This can be explained as follows: (1) the probability that the path is passable, $p(\theta_1)$, is much higher than the probability that the path is impassable, $p(\theta_2)$; (2) the loss incurred by taking the alternative path, $L(\theta_1, a_2)$, is higher than the loss incurred by the deliberative manoeuvre, $L(\theta_1, a_1)$; (3) The sonar information is not certain. Note that although the robot has to backtrack and incurs a certain loss in this traversal, it however learns something new about the state θ. The blockage information is sent to the global path planner to increase the uncertainty cost for this portion of the path to take into account when planning an global optimal path next time.

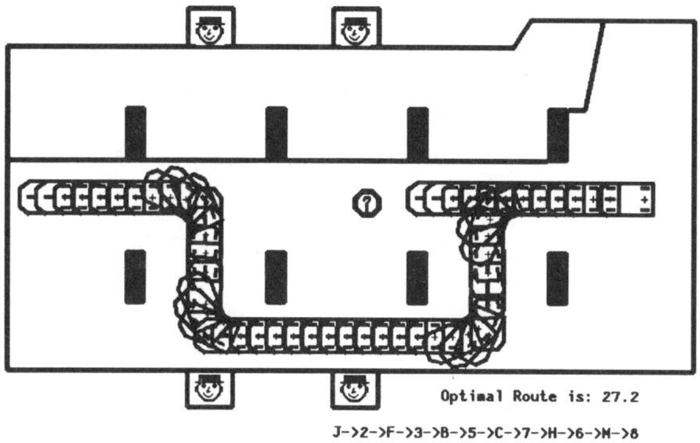

Figure 4.12: An action *backtrack* which has minimum expected loss

Following the examples described above, it is clear that the Bayes rule we seek makes maximum use of the *a priori* information about the state θ, i.e. $p(\theta)$. If the prior distribution $p(\theta)$ were changed sufficiently, the Bayes rule would change. However, in our application, this prior information for current decision-making is actually a posterior distribution which is combined with all knowledge that is obtained so far by previous traversals. It will be more precise to inference the true state θ as the robot moves around in its environment.

4.6 Real-time Implementation

Figures 4.13 and 4.14 show the results implemented by the mobile robot in our AGV Laboratory. The image sequences in the two figures demonstrate how the robot handles an unexpected obstacle, a chair in this case, by taking either a *manoeuvre* or an alternative path towards the goal position that has been commanded. The robot speed is 0.6m/s. There is no stop during maneouvres.

Figure 4.13 consists of a sequence of eight frames, from frame 1 to frame 8. It clearly shows that the mobile robot takes an action to sidestep a chair which partially blocks the path. This corresponds to Example 1 described above.

Figure 4.14 shows that the mobile robot detects an unexpected chair when it moves towards the goal which is at the end of the corridor. After taking extra observations at close range, the robot found the path is impassable. Then it takes the alternative path and reaches the goal position eventually. This corresponds to Example 2.

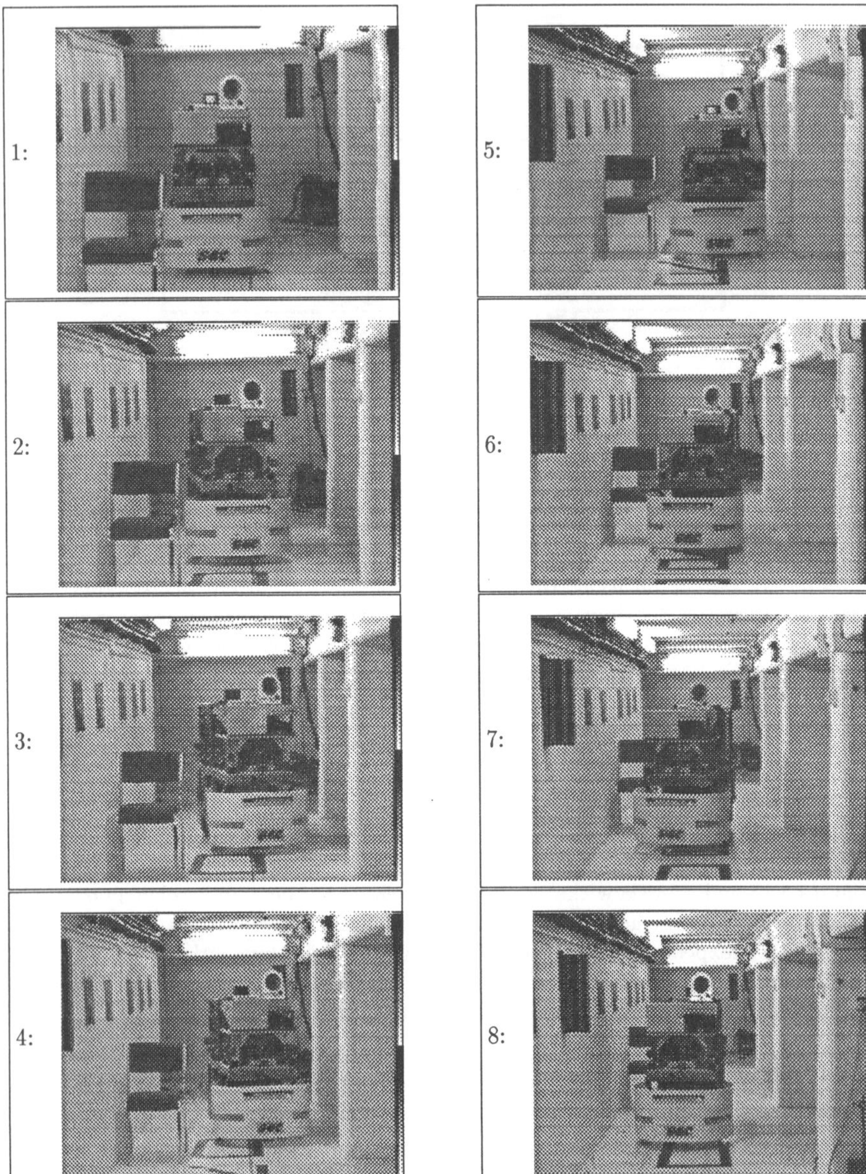

Figure 4.13: The AGV sidesteps to avoid an obstacle

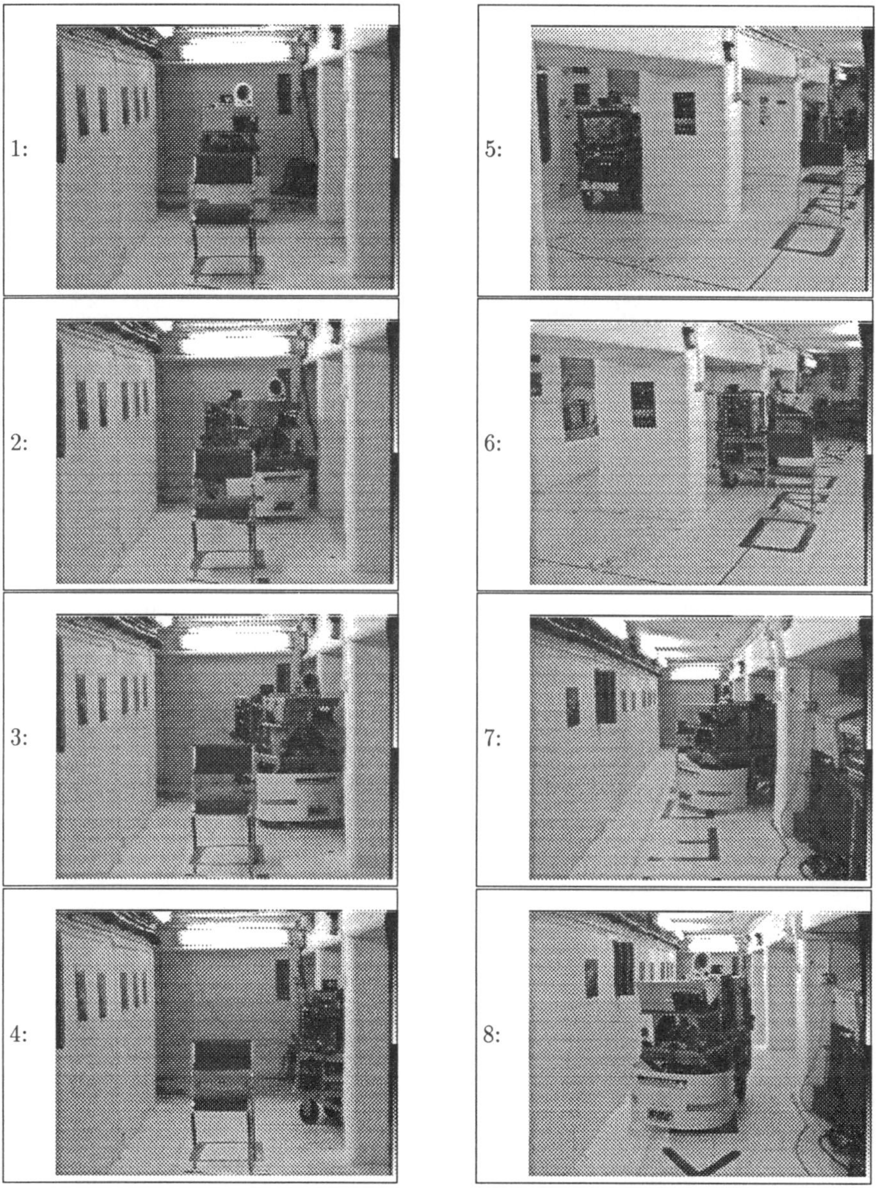

Figure 4.14: The AGV avoids an unexpected obstacle

4.7 Conclusions

In this chapter, we have formulated a framework for global planning and obstacle avoidance using decision theory. The solution is not only collision free but also optimal in terms of time cost, taking into account the constraints imposed by the roadway and the vehicle kinematics. An attempt is made to provide a formal foundation for local path planning under uncertainty using Bayes analysis and decision theory. The major contribution is the treatment of uncertainties introduced by both the sensing process and the unexpected obstacles. The proposed strategy has been implemented by both simulation and on a mobile robot in real-time.

Part II

SENSING AND NAVIGATION

Chapter 5

Low Cost Range Sensors for Reactive Planning

Penelope Probert

In the real world, a robot needs to be able to detect and handle changes. A cleaner may drop a bucket in a factory, a suitcase may be in the way in a baggage handling hall, a previous robot may drop its payload. Currently many autonomous vehicles have fixed functionality and cannot cope with such situations; often indeed they cannot even detect them except through bumpers. Future generations need the ability to observe the environment and to replan. The Oxford project aims to make vehicles react to the environment through dynamic planning based on fast sensing. The realisation of reactive planning depends on architecture design (see for example work by Brooks, Payton and Arkin amongst others) [34, 157, 9]. However its success is fundamentally limited by the reliability and quality of sensor data.

The choice of sensor for reactive control is dictated by the ease with which processing can be directed immediately towards local planning. Reactive planning needs high bandwidth sensors. Factory vehicles typically move at speeds of between 1 and 3 m/sec (1 m/sec is a fast walking pace). If replanning is required at, say, 5cm intervals, a bandwidth of between 20 and 60 Hz is required. These bandwidths may be sufficient for a road vehicle too since the road environment should be less cluttered. Such bandwidths, which must allow both for data acquisition and processing, are hard to achieve reliably at low cost.

Passive vision is a popular sensor and has received a great deal of attention in robotics. With specially built hardware and a suitably controlled environment, real time performance is possible. Graefe and others [169, 80] are developing a road following vehicle with camera control which is to be capable of driving on the German autobahns. Demonstrations on empty roads at speeds of 14m/sec have already been achieved, and work is now proceeding too on the detection of other vehicles. High frame rates [79]) have been achieved through careful modelling, well designed algorithms and the use of appropriate parallel processing. Inaba and Inoue [105] have built and demonstrated a multi windowing camera to track moving objects, again through careful design of algorithms and hardware. In a less serious but equally demanding application, Andersson [8] has demonstrated a robot ping-pong player which reacts within 0.4-0.8 seconds. In chapter 8, Wang, Brady and Bowman describe work in real time vision in the Oxford AGV project.

Figure 5.1: The AGV equipped with a line camera, 3 sonar sensors, the NEL laser scanner, and a sensor to detect movement of the front bumper

These systems rely on well understood environments, and on special purpose hardware to run efficient algorithms. An alternative way to provide high throughput rates is to reduce the amount of data. In Oxford we have developed a feature tracker using a line CCD camera and rely on the fast frame rate to make up for the loss of spatial information. Line cameras are not suitable for totally unstructured environments but our work [161] has shown that they can handle environments far more complex than their traditional applications in inspection and line following robots. Brooks [34] has implemented a stereo pair of line cameras, using cylindrical lenses to widen the vertical field.

In uncontrolled environments, it is not feasible to exploit modelling to the same degree. For low cost, on-board systems direct measurement of range using *active* sensors is a more practical solution. Range images are not as sensitive to prevailing light conditions as passive systems and image interpretation is usually easier; the disadvantage is the need to include an active device in the sensor. The need for a high throughput rate means that we may have to rely on two dimensional information on the environment; for example range data acquired by scanning a spot. We aim to compensate for the loss of three dimensional information by acquiring images rapidly over time.

A number of range sensors are used commonly, the most popular being sonar. Optoelectronic sensors are now comparable in price and in many environments provide data of much higher quality. Other sensors, such as radar and lidar (laser based radar) are currently much more expensive and the large range they provide does not normally justify the cost in a factory. Nevertheless millimetre wave radar is becoming of increasing interest and can be realised in integrated circuit technology, leading to compact sensors. The wavelength is similar to sonar.

In the first part of this chapter we present a general treatment of range sensing, bringing together the basic equations and the causes of error. We then describe current work in the two most popular technologies: sonar and optoelectronic sensors in more detail. Other chapters in the book describe these sensors further in the context of developments within Oxford.

5.1 System Principles

Range sensors rely on at least a portion of the light emitted being scattered back towards the receiver. In this section we describe the effects of signal scattering, target type and the transmitter/receiver design on their accuracy and application. We discuss how the performance characteristics are related to the transmitter wavelength.

5.1.1 Attenuation in the Path

We start by looking generally at the variation of signal with path length. All the ranging sensors we consider propagate a wave, electromagnetic or sonar, to a source and detect the wave scattered from the source. The path attenuation can be divided into parts: the attenuation owing to beam spread and the attenuation added by losses in the medium.

Consider first the attenuation from beam spreading in a divergent beam, typical of radar or sonar. At a target, the power density, P_A, is given in terms of the transmitter power P_T and the range R by the expression:

$$P_A = \frac{GP_T}{4\pi R^2} \tag{5.1}$$

where G is the gain or directivity of the antenna, which is defined in terms of the power radiated in a given direction compared with the isotropic power [180].

The outward wave is reflected from the target and a receiver detects power scattered isotropically from the reflecting surface. For isotropic scattering, each point on the scattered wavefront is again attenuated by a factor of $(4\pi R^2)$. Defining an effective aperture A which describes the collecting area of the receiving antenna, and a target scattering coefficient, σ, the power received, P_r from the target is given by the expression:

$$P_r = P_T \sigma G A (\frac{1}{4\pi R^2})^2 \tag{5.2}$$

For a single antenna, G and A are related through the wavelength λ, and so we can write:

$$G = \frac{4\pi A}{\lambda^2}; \quad \text{hence:} P_r = \frac{\sigma}{4\pi} P_T (\frac{G\lambda}{4\pi R^2})^2 = \frac{\sigma}{4\pi} P_T (\frac{A}{\lambda R^2})^2 \tag{5.3}$$

This equation is sometimes known as the *radar* equation. We note that even in a lossless medium, the losses due to scattering increase with R^4; therefore the signal to noise ratio at the receiver falls off sharply with range. Other conventions exist for defining the scattering coefficient; here we follow the the convention for radar [180]. The scattering coefficient is hard to assess, depending on texture and shape.

However most sources are well focussed, and we can model the emitter as a collimated beam (for a laser source this is of course especially appropriate). Then there are no spreading losses on the outward path and in the ideal case we assume that the target receives the full incident power. The path attenuation then only increases as R^2.

Sometimes retroreflective targets may be appropriate. Retroreflective targets are shaped so that all incident power is reflected back along the incident axis or in a given direction, so there is almost no R^2 attenuation. Wavefront distortion may be very low, only a few wavelengths. Experiments with the laser bar code scanner on the AGV (chapter 2) show that the retroreflective bar codes can be read at ranges up to about 50m.

As well as scattering losses, attenuation arises from losses in the transmission medium. Figure 5.2 (from [185]) and Figure 5.3 show the attenuation characteristics for electromagnetic radiation in the atmosphere, and the effects of water vapour. Figure 5.4 (from [126]) shows similar characteristics for ultrasound waves.

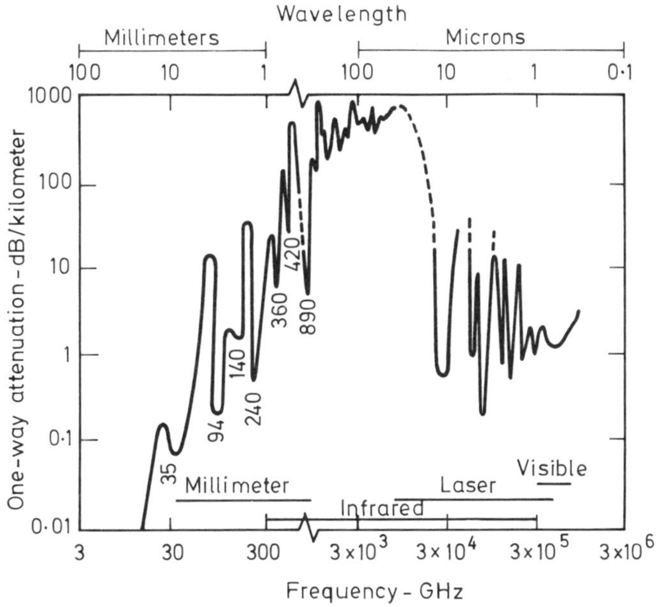

Figure 5.2: Attenuation characteristics of electromagnetic radiation in the atmosphere. Military millimetre wave radar uses regions around 94GHz

5.1.2 Selection of Wavelength and Radiation Mode

One of the first steps in choosing a sensor is the choice of wavelength: microwave, millimetre wave or optical. In the millimetre wave region there is also the choice of modality: ultrasound or radar. It is clear from Figure 5.3 that the wavelength affects attenuation characteristics. The wavelength and the mode of radiation (electromagnetic or sonar) also affects the operation of the sensor in other ways.

Antenna directivity

The most obvious effect of wavelength is on the antenna gain and aperture. For high gain and effective aperture electromagnetic theory shows that the antenna should be large compared with wavelength. For example for a parabolic reflector of diameter D most energy is confined mainly to a solid angle λ^2/D [143]. It is therefore much easier to focus a beam at small wavelengths than at long ones: for example an optical beam can in theory be focussed within the device itself whereas millimetre waves and sonar need larger external focussing lenses. Microwave radar, which has wavelengths of several centimetres, needs relatively large dishes to produce high gain - think of satellite TV dishes. An alternative to using a single antenna is to use a phased array,

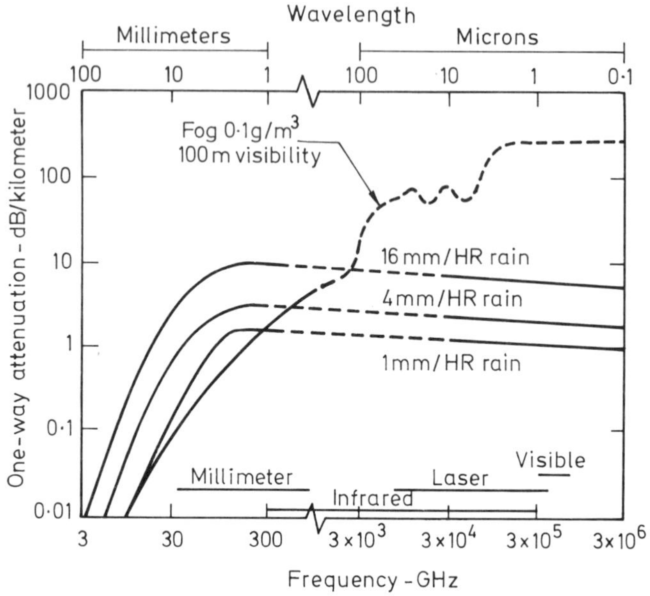

Figure 5.3: Effect of water vapour on the attenuation characteristics of electromagnetic radiation in the atmosphere

which provides a narrow beam. If the transmitter is steered electronically carefully designed feeds are needed; a simpler approach is to steer the receiver through software.

Scattering characteristics of target

We know that different targets react differently to different wavelengths: a dark target does not reflect much light but reflects sound waves; an acoustic baffle does not reflect sound but may reflect visible light. Particular applications may also need special sensors. For example sensors in the visible spectrum are poor in smoky environments whereas sonar is not attenuated so significantly (although ultrasound waves are bent if there are temperature gradients owing to the dependence of the velocity of sound with temperature). At optical frequencies, ultra-violet reflects better from dark surfaces than infra-red.

The texture of the target is also an important feature since it determines the type of scattering and reflection. In general the scattering is more uniform with a coherent source. However the main problem arises when the wavelength is comparable with the granularity of a structure. Then the scattering is no longer Lambertian but instead specular reflection occurs: the incident wave is reflected as from a smooth surface.

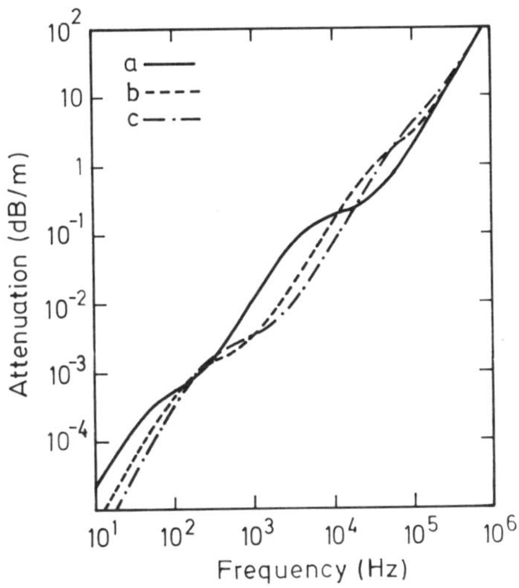

Figure 5.4: Attenuation characteristics of ultrasound radiation as a function of frequency at 20° in air the atmosphere. (a) for 10% relative humidity, (b) 40% relative humidity (c) 80% relative humidity

Specularity is a particular difficulty in using sensors with relatively long wavelengths, such as sonar and millimetre wave radar. Indeed Brown [37] suggests that nearly all sonar reflections are specular. Quoting from [37]: *Using ultrasonic sensors to look at arbitrary objects in a room is rather like standing in a room completely filled with mirrored objects and having only a penlight glued to your forehead as a source of light.* Certainly raw sonar range maps show little resemblance to reality. The development of methods to interpret sonar data reliably has been a triumph of careful modelling and analysis.

Specularity depends primarily on the wavelength, not the type of wave. Standard wavelengths for ultrasound are between 3 and 6mm (50-100kHz). A standard wavelength for millimetre wave radar is 3mm (frequency of 94GHz) as this corresponds to a window in the attenuation characteristics (Figure 5.2). Therefore millimetre wave radar suffers from specularity too. The advantages of radar are much longer range and negliglible time of flight; the disadvantage is cost. However the cost is becoming low enough for application in mobile robotics to be reported [178].

5.2 Sensor Configurations

The sensor's performance depends on the detection method as well as the source. We can categorise sensors by the method of detection:

Direct time of flight

This is the commonest method of detection for sonar, in which the time of flight is generally of the order of 3–50 milliseconds depending on range. The signal is transmitted as a pulse and received some time later, usually using the same transducer. The start pulse of an electronic counter is synchronised to the start of the transmitted pulse and detection of a returned pulse above a threshold amplitude stops the count. A direct read out then provides the time of flight.

Amplitude detection

The simplest method of detection. By itself amplitude does not give range information without extensive calibration and knowledge of the environment; however it can be used with other techniques to assist in feature detection. There has been a small amount of work in measuring in the time domain the amplitude of the echo return in sonar to determine both near and further reflections, but the data is hard to interpret without a narrow beamwidth.

Time of flight from phase detection

This method is becoming popular for optoelectronic range sensors. The transmitter (light emitting diode or laser) is modulated by a reference radio frequency signal. The radio frequency signal received is mixed with the reference at the receiver and the time of flight determined from the phase difference. We will refer to this technique as *amplitude modulated continuous wave (AMCW)* detection.

Frequency modulation or chirping

The frequency of the transmitted signal is modulated by a triangular wave. On its return it is mixed with the current outgoing signal. The difference frequency depends on the time difference between the original and current transmissions. The technique is standard in radar and in the more sophisticated laser radar sensors. It has been applied to ultrasound sensors mainly in the medical field. Kay has developed a number of sensors for partially sighted people which provide a tone related to the distance of the nearest object. With training these can be very useful; a stereo pair of these sonar sensors has allowed a blind person to play Frisbee!

A frequency modulated radar signal from an outdoor scene (a container port) is shown in Figure 5.5, together with its Fourier transform which shows which frequencies are present, and hence at which ranges reflections occurred. You can see clearly two distinct objects (a person and a building).

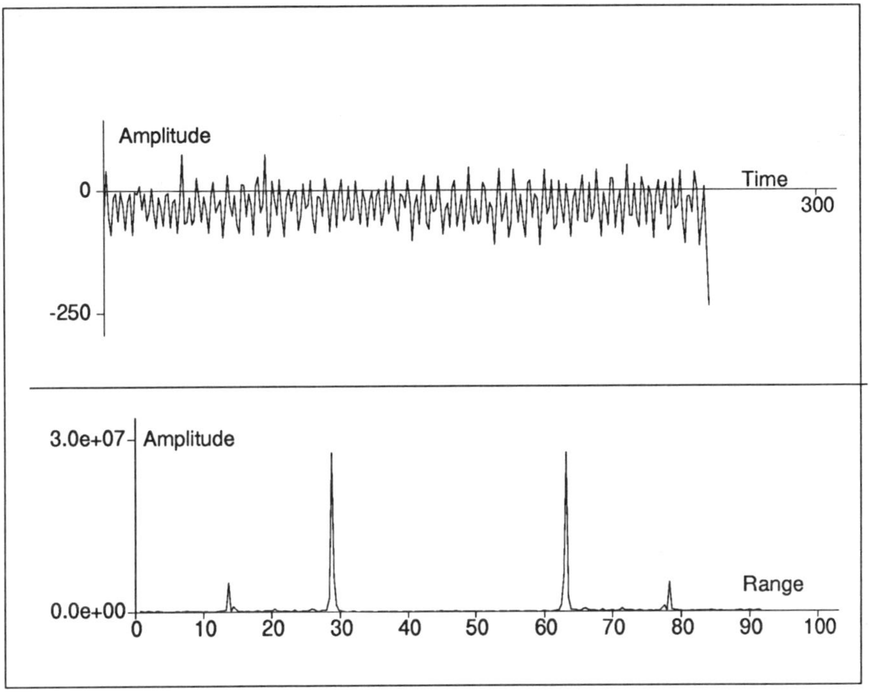

Figure 5.5: A frequency modulated radar signal. The top trace shows the signal returning and the lower one its Fourier transform calibrated in range (image provided by Hugh Durrant-Whyte)

Triangulation

These are built on a geometry which converts a range reading into a horizontal offset on a lateral detector, usually using a laser source and a CCD detector. High accuracy can be achieved over a small depth of field. Triangulation sensors are described in detail in chapter 6.

5.3 Choice of Sensor

The last sections have described the limitations and general principles of active range sensing. Which sensors are the most promising?

In practice two factors have led the selection of robotics sensors. First, most of the initial impetus in autonomous robots was by workers in computer science and artificial intelligence, who had little interest in hardware development. Secondly, robotics is not a large enough field to support a mass market in cheap sensors. As a result many of the popular sensor technologies use configurations which are designed for quite different applications, so the technology performs well below its capability in robotics. This weakness has been turned to a strength in leading the development of serious sensor modelling techniques. However the future lies in exploiting hardware technologies as well as software models.

Until recently nearly all mobile robots used sonar, following technologies adopted for the consumer market. Nearly all robotics researchers have used a sonar sensor developed by Polaroid (figure 5.6). The Polaroid sensor was designed for application in cameras and, not surprisingly, leaves much to be desired as a robot sensor. In general sonar sensors work well as a bumper but a lot of processing is needed if they are to direct more detailed navigation. Optoelectronic sensors also rely on technologies developed for other industries, especially on the low cost high powered light emitting diodes and laser diodes developed for the communication industry (and the CCD line arrays used for fax machines in triangulation sensors). The more expensive contender is millimetre wave radar which has been developed to a very high degree of sophistication in the military sphere.

Each sensor technology has its good and bad points. We summarise the major features of each type of sensor:

- Range. Cheap sonar sensors measure ranges from about 5cm to 10m using direct time of flight measurement. The range of optical sensors based on optoelectronic technology is comparable, but extends down to very close ranges. Millimetre wave radar is suitable for much longer ranges.

- Accuracy. In range, an accuracy of about 1-2cm is usually sufficient and can be achieved by all three technologies. The distribution of error depends on the technique for determining range. In direct time of flight measurements the error is approximately uniform over the full range, with rapid deterioration when the

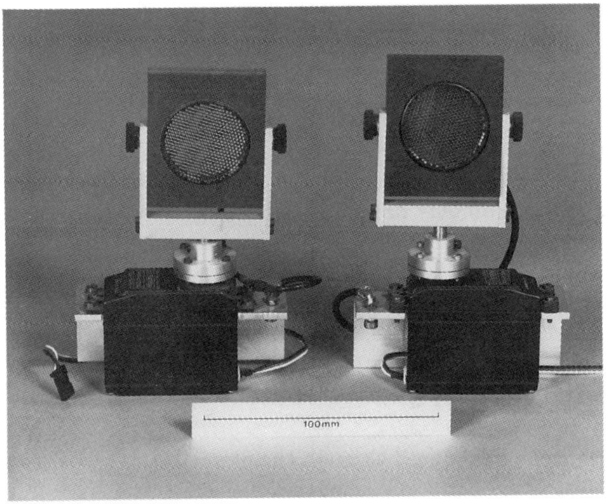

Figure 5.6: Two Polaroid sensors from the AGV sonar array

signal to noise ratio drops below a certain threshold. However most analogue techniques for range detection give better accuracy at close range, since the accuracy depends on the signal to noise ratio. Good angular resolution is easiest to achieve with optoelectronic sensors because of the small wavelength.

- Cost. For commercial acceptance, the cost of sensing should be a small fraction of the cost of the vehicle. Sonar sensors and simple optoelectronic sensors can be built for a few tens of dollars and so have commercial application in low cost systems, and in industries which are fiercely competitive, such as the automobile industry (for example sonar sensors to assist in parking are built into some of the higher range models of automobile). Millimetre wave and laser radar are used in outdoor vehicles following developments mainly in the military sphere where range is more important than cost. However modern fabrication technologies are such that a millimetre wave radar system could be purchased for a few thousand dollars if a bulk market were established.

- Specific environmental considerations. Optoelectronic sensors fail in dark, matt environments whereas sonar fails in acoustically dead environments. All three technologies (and indeed any active sensing technology) fail under specular conditions; sonar and millimetre wave fail much more frequently than optoelectronic sensors because of their longer wavelength. Conditions such as mist and rain also limit the frequencies which should be used outdoors.

- Size of the sensors. The beamwidth depends on the size of the focussing antenna relative to wavelength. Optoelectronic sensors are the easiest to focus.

For the best results, most mobile robots use a number of sensors, integrating their information to optimise the sensing strategy.

In the rest of this chapter we describe the application of the low cost technologies which are used most commonly in reactive planning. We start with the commonest sensor, sonar, looking at recent successes in planning and tracking. We then describe some optoelectronic sensors, especially AMCW sensors. We leave triangulation sensors for Chapter 6.

5.4 Sonar Sensors

Ultrasound signals can be induced through the piezoelectric effect or through electrostatic forces. Most sensors used in robotics are electrostatic since this mechanism is more efficient for coupling into air. Typical frequencies are between 40 and 100kHz, the higher frequencies being easier to focus but suffering greater attenuation. The same transducer can be used for transmission and reception.

Ultrasound transducers can be obtained for a few pounds. The commonest type reported in robotics is manufactured by Polaroid. It can be used to measure ranges from about 0.25m to 10m. A special foil, insulated on one side, is stretched over a grooved metallic plate. The back of the plate in contact with the insulating side of the foil forms a capacitance which, when charged, exerts an electrostatic force on the foil. Application of a radio frequency wave to the backplane therefore induces longitudinal waves. Similarly, the foil can be used to convert longitudinal waves to a radio frequency voltage in reception. The construction is shown in Figure 5.7.

Range is determined through direct time of flight measurement. A firing pulse triggers an ultrasound burst from the sensor and starts a counter. The counter is halted when the sensor, now acting as a receiver, detects a signal above a pre-set threshold; the counter reading gives the time of flight. A significant advantage in the Polaroid transducer is that all the analogue electronics can be purchased, so only timing circuits need be built.

The diameter of the transducer determins the beam pattern. As shown in Figure 5.8 power is transmitted over quite a wide angle. The rather large beamwidth makes the transducer excellent as the basis of a sonar bumper.

Figure 5.9 shows a scan around a room using this sensor. We note the following two points:

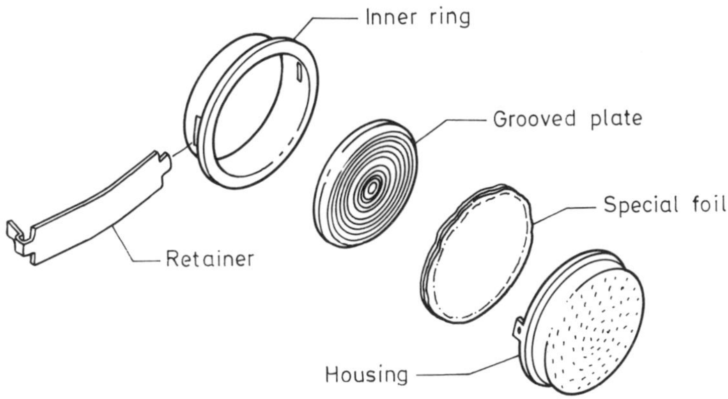

Figure 5.7: The construction of the Polaroid sensor

1. The data bears little resemblance to the physical room layout. This is mainly because of specular reflections; also because the range recorded is that from the nearest object within the beam which may lie some way off centre. Feature detection on raw data is not feasible.

2. Features are distorted, in particular straight edges appear curved. We can explain the distortion as follows. The unit always returns the time at which the first part of the signal is returned. When the beam is normal to a surface the earliest reflection to return is from the centre of the beam since this travels the shortest path length, and is indeed the range required. However as the transducer rotates the shortest path is associated with different points on the beam and will be less than the path travelled by the beam centre. Therefore curvature results.

The least processing needed for sonar data is the removal of specular reflections. The simplest way to do this is simply by a time out in hardware: if no signal is received within a given interval a *no reflection* is registered. This does not necessarily mean that there is nothing there; it is simply uncertain data and must be used in conjunction with other data to make a final decision.

5.4.1 Application in Reactive Planning

The most notable success story in ultrasound comes from the natural world, in the abilities of bats to echo locate. Far too little is understood still about the natural world; we are unable to match the capabilities of insects in handling the unexpected,

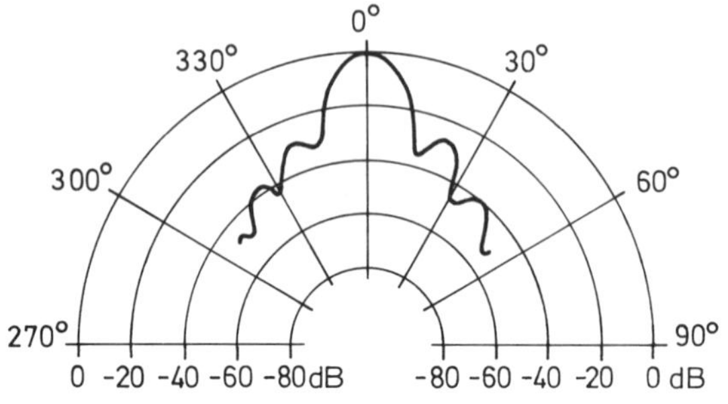

Figure 5.8: Beam pattern of the Polaroid transducer

let alone bats. It is not only the reliability of the sensing which is so impressive, but also the speed. Kuc and Barshan describe an attempt to mimic bats' navigation through echo-location in [14].

The simplest use of sonar in mobile robotics is as a bumper. The wide beamwidth is advantageous in providing 360° coverage. Assuming a beamwidth of 15°, taken typically for the Polaroid sensors, 24 sensors provide full 360° coverage. However a smaller number may be acceptable, with the advantage of reducing crosstalk between adjacent sensors. Hu and Probert [167] give a typical example of a sonar bumper in obstacle avoidance using 12 Polaroid sensors mounted around a vehicle. Chapter 4 describes a decision theoretic approach to obstacle avoidance using the bumper.

More sophisticated use of a sonar array in planning in a field of obstacles is described in a number of papers by Borenstein and Koren [24, 27, 28]. The hardware configuration is similar with Polaroid sensors in a ring around a vehicle, but the sensors drive path planning and mapping algorithms. A map is held representing the environment as a regular grid, with a number associated with each grid square representing the probability of the presence of an object in that square (a representation often called an uncertainty or occupancy grid) [60, 56]. The map is combined with modified potential field path planning (which the authors call a virtual force field). Potential field planning is described in Chapter 11.

To build up a good occupancy grid, a large number of readings have to be taken from different robot positions to eliminate the effects of specularity and noise. However in reactive planning there is only time for the robot to take readings in one or a few close positions. Averaging a number of readings taken from a single position reduces random noise, but errors from specularity are systematic and will be reinforced through averaging techniques. Specularity therefore remain a problem.

A further problem results in an array of sonar sensors; crosstalk between neighbouring sensors. This is commonly eliminated by firing only a few sensors at a time

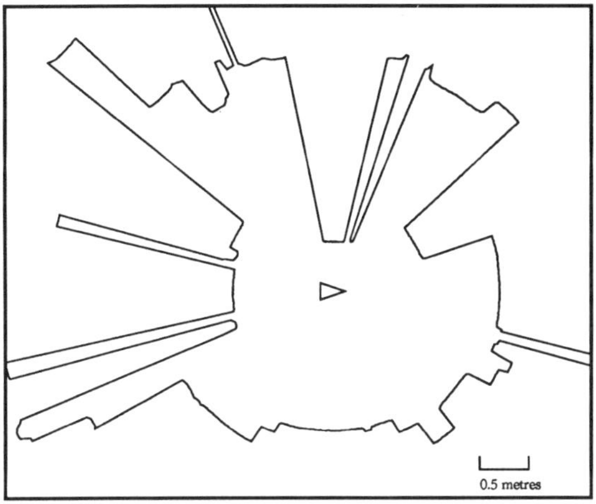

Figure 5.9: A typical scan taken with a Polaroid sensor in the AGV laboratory in Oxford (image provided by John Leonard)

(for example those facing at 90° from each other) and waiting for any echoes to die away before firing the next set. However this is slow. Borenstein and Koren [28] propose a particular firing sequence to eliminate crosstalk. The method uses different delays between firing sets of sensors, alternating between long and short delays, and removes crosstalk by comparing consecutive readings. With the elimination of crosstalk, the rate of firing depends only on the maximum distance to be measured; there is no need to allow for multiple reflections to die away before firing the sensors again.

Using this method, the authors claim to drive a robot at 1m/sec through a field of densely spaced, pencil thin poles (note that narrow cylinders always return some reflected power along the line of incidence; hence the environment is essentially non-specular). The system is robust to external noise too; for example to another robot with sensors firing at the same frequencies. It is an impressive demonstration of the power of sonar, although in a less controlled environment specularity would be a problem.

5.4.2 Advances through Modelling

These applications use sonar to give a coarsely discretised view of the local environment. However the robot may need accurate information, for example to locate a narrow doorway. We have seen several major problems with using sonar for such accurate measurements. First, there are frequent misreadings from noise, other sensors or peripheral objects such as the floor. Second, there is poor directionality. Third, there is the problem of specularity.

Borenstein and Koren's approach remedied the problems caused by random noise and interference. They made no attempt to handle the problems of specularity and poor directionality. However Brown [37] showed that, rather than being a problem, specularity and beamwidth could be exploited. He realised that a sensor always returned the value of the normal distance to a surface, provided a portion of the beamwidth was incident normally on the surface. Any portion of the beam which was not close to normal incidence was specularly reflected. Using this model, he showed how range measurements from a number of sonar sensors could determine shape as well as range.

The essential specularity of ultrasound in air was also exploited by Kuc and Siegel [124] in a paper leading the way to the efficient simulation of sonar transducers. The authors analyse the impulse reponse of various features (a wall, a corner, an edge) as a function of transducer size, location and orientation. Their work is picked up later in the development of small sonar arrays. With Barshan, Kuc has developed a sensor pair (with two receiver/transmitters firing synchronously) capable of distinguishing between corners, planes and edges [13]. The maximum distance between transducers is determined by the need for the beams to overlap on targets; the minimum distance is determined by the need to discriminate between the responses given sensor noise.

Using similar principles, Leonard and Durrant Whyte [133] have demonstrated the application of ultrasound sensors in accurate localisation. Arguing that a natural representation for a sonar scan would yield best results, they introduce the idea of *regions of constant depth* for representing scan data (see chapter 7). They used the sonar signature to build up a map as the robot moved around, and then again for localisation within that map. Accuracies of better than a centimetre resulted, an impressive result given the poor directivity in the Polaroid sensor. However a large number of scans had to be taken, and so the techniques was slow to run, limited by the speed of sound. The work is now being continued using 2-element arrays leading to application in real time, as described in chapter 7.

The simple two element arrays point the way to serious development of ultrasound arrays for beam shaping and steering. For versatility and especially control of beamwidth the transducer should be integrated into the array itself. Phased arrays can meet a wide range of specifications. In the medical field there has been long standing work in this area, for example see [138]. One of the first to develop sonar arrays for use in air was Kay [114]. The geometry and control can be developed to service diverse applications. For example at Nottingham University, UK, a number of phased arrays have been developed, with beamwidths varying from 2.5^o (for pick and place operations) to 60° (for general guidance) [148]. The arrays uses a single transmitter and software to realise the phase pattern in the receiver, using a digital signal processor chip for real time performance.

5.5 Low Cost Optoelectronic Sensors

Off the shelf ultrasound sensors are easily available and of low cost, but, as we have just seen, the data is very hard to interpret. Most of their difficulties are associated with the long wavelength, and we look to optical sensors for greater ease of use. Optoelectronic components are of low cost for application over moderate ranges, where transmitters of a few to tens of milliwatts can be used. The main cost of the system is the optical configuration to focus and direct the beam, but we see later in this section that for AMCW sensors we can rely on processing to remove most problems associated with poor focussing. The price of optoelectronic rangefinders varies from tens to thousands of dollars; in line with the philosophy of low cost sensing we look at the cheaper end of this spectrum.

Optoelectronic sensors use either a laser or a light emitting diode as the source. Laser diodes are normally used rather than gas lasers. Lasers emit collimated single frequency light and are used for accurate probing of the environment. LEDs are more suitable for less accurate measurements. Both may need external focussing lenses, and we have had the best results with standard commercial camera optics. LEDs are unconditionally eye safe and easier to drive; therefore for many applications they are the obvious choice. The receiver is typically a photodiode or, if greater sensitivity is needed (at the expense of high bias voltage), an avalanche photodiode.

Figure 5.10: GEC infra-red bar code scanner used for localisation in the Oxford AGV

5.5.1 *Sensors Based on Amplitude Detection*

The simplest sensors to build detect the amplitude of a returned signal. Amplitude detection alone is a poor method of measuring range, since the amplitude depends on target reflectivity and texture as well as range. Texture is the more important; Everett and Flynn [63] show that in the near infra-red (880nm) the variation in reflectivity with surface colour is only 15% but there is very considerable variation with texture. Similar differences arise in other bands of the optical spectrum.

One way of using amplitude detection is through suitable modification to the environment. For example the bar code scanner on the AGV (figure 5.10) is simply an edge detector; the bar codes are designed so that each presents a different edge pattern. Knowledge of the position of each bar code allows the robot to determine its position to an accuracy of better than 1cm.

Everett and Flynn [63] describe a programmable near infra-red amplitude detection sensor for navigation in an unstructured environment. Each sensor consisted of four light emitting diodes (LEDs). Between one and four light emitting diodes can be fired, depending on the range expected, and the number of diodes fired before a signal was detected gave a very crude estimate of range. A robot was equipped with 35 such sensors, which measured distances up to about 4.5 metres. In spite of the very crude range measurement, discontinuities in the amplitude received gave a very exact estimate of feature edges.

Infra-red amplitude detection is more useful in conjunction with other sensors. Flynn later used the sensors with sonar to detect the edges of doorways [69]. A similar sensor combination is used by Wen [195] to locate holes for docking container vehicles. Amplitude data may also be useful in segmenting range images.

5.5.2 AMCW Sensors: Ranging through Phase Detection

Recent developments in low cost range detection have concentrated on using phase detection on amplitude modulated carriers. The optical device, light emitting diode or laser diode, is modulated at a radio frequency of a few MHz. The path length is determined in terms of the phase difference between the outgoing and incoming signals at the modulation frequency.

There is an inherent ambiguity of π in standard phase detection methods, and this limits the maximum range of the sensor to one quarter of the wavelength of the modulation signal (the distance to the target and back must be less than one half of a wavelength). This limits the modulation frequency to a maximum: for example a typical modulation frequency of about 5MHz would imply a maximum range of 15m. Given this limitation high modulation frequencies are generally advantageous (at least until special high frequency circuit techniques are required) in order to maximise the phase shift for a given distance.

The phase shift need not be measured at the modulating frequency since phase is preserved over frequency conversion. Rather than measuring phase differences which might correspond to picoseconds in time at a high frequency, modulation to a lower frequency in the receiver allows us to measure the same phase at a lower frequency where the time equivalent is much longer. Therefore we can use a high frequency signal in transmission to maximise the phase difference for a given range but work at low frequencies for greater accuracy in the detection process. The catch comes in

Figure 5.11: An AMCW sensor used in Oxford

the mixing stage since the local oscillator must be stable in amplitude and phase; a combination hard to achieve for high frequency signals.

Sensors scanning in one and two dimensions (i.e., which provide two and three dimensional images repectively) have been described in the robotics literature and are becoming available commercially. Miller and Wagner use a 2mW light emitting diode and a photodiode receiver [142], which is 100% modulated at 5MHz. The rangefinder uses a Fresnel lens to focus incoming reflected light and a rotating mirror to provide 360° scanning coverage up to about 6m; the cost of parts is less than $700 (1988 prices). The sensor was used extensively by Cox in map building and planning [54]. There is no superheterodyne stage in the receiver; all phase measurements are made at the transmission frequency of 5MHz. A best accuracy of about 2cm is achieved at a 1kHz sampling frequency. Adams and Probert have modified a version of this sensor for use in reactive control and local path planning (see figure 5.11)[3]. A scan in the Oxford AGV lab is shown in Figure 5.12. This scan was calibrated using a polynomial correction; other designs remove the need for calibration (see Figure 5.13). Compare the quality of these figures with the sonar scan in Figure 5.9.

There are a few reports of work using similar sensors. For example Singh and Keller use two dimensional data from an AMCW rangefinder [183] for high speed autonomous navigation, using a frame rate of eight lines (1000 points per line over 180°) per second. Hebert, Krotkov and Kanade [91] use a three dimensional AMCW sensor working as the main component of a perception system for a planetary explorer. Once the extra dimension is added, processing becomes a significant issue especially in terms of time. This system is more powerful and expensive than the others we have described, consisting of a 100mW laser diode operating at 820nm and scanning using vertical and horizontal mirrors.

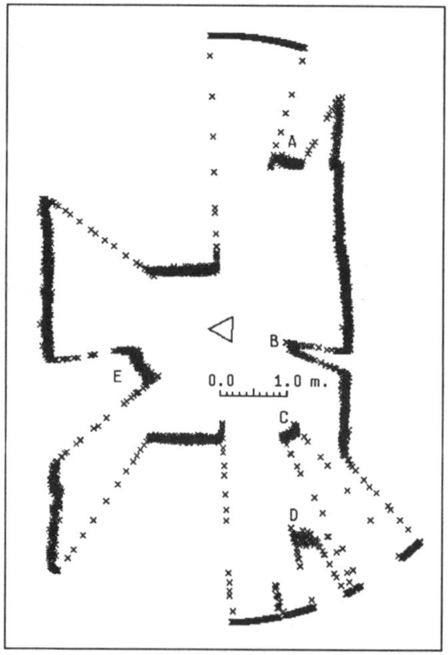

Figure 5.12: A data scan from Oxford's AMCW sensor in in a relatively cluttered environment. In the upper picture the labelled points correspond to: A: a cardboard box; B: a circular cross sectioned fire extinguisher; C: a battery charger; D: the AGV; and E: Oxford's Robuter vehicle

With good design the repeatability of AMCW sensors is comparable to triangulation sensors, and better at the longer ranges. Hofler [95] decribes an AMCW sensor designed for accurate inspection. This sensor has a frequency conversion stage in the receiver and, with a 30mW laser, achieves a repeatablity of better than 1mm

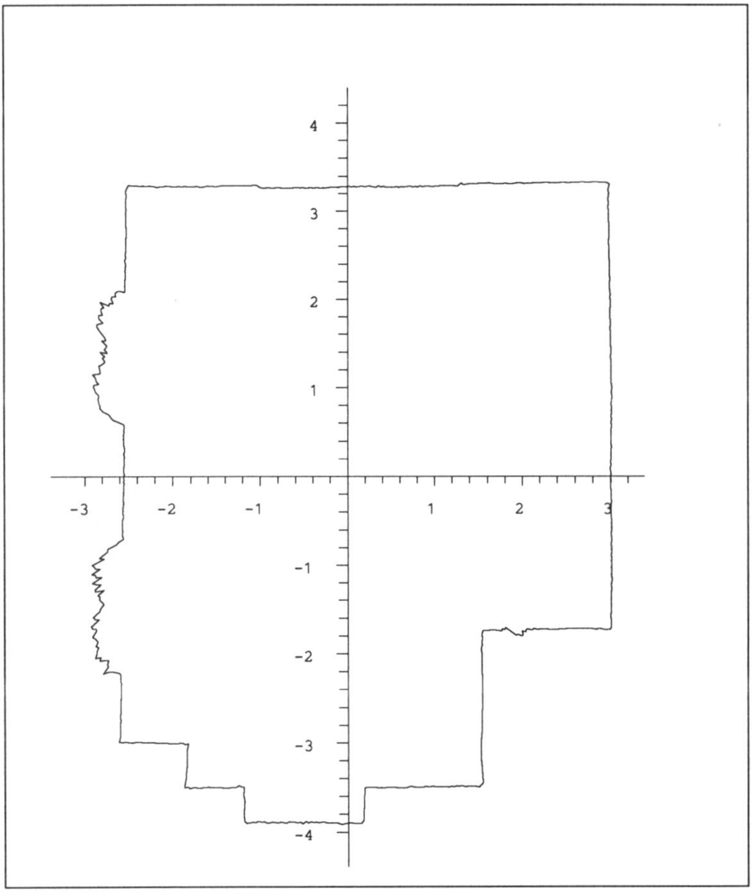

Figure 5.13: Raw range data from an AMCW sensor designed by Brownlow[38] illustrating the excellent linear characteristic (The scene is of an irregular-shaped room, with curtaianed windows on the left-hand side)

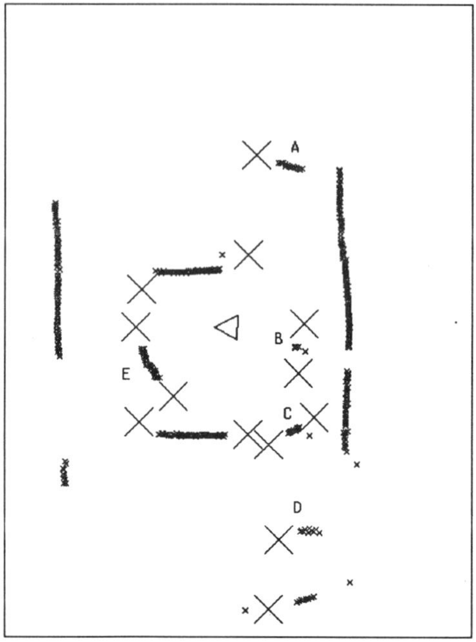

Figure 5.14: The results from processing the scan shown in Figure 5.12 with an algorithm which determines possible targets (X) for the robot to move towards in free space. The algorithm has three stages: the removal of mixed pixel points, the recognition of edges through running a Kalman filter to match line segments and the application of two simple rules to determine the extent of free space around each edge [2]

at ranges up to 1m and a sampling rate of 1MHz. These figures are probably close to the ultimate accuracy of the technique. An avalanche photodiode is used as the detector to increase sensitivity. Brownlow [38] also uses an avalanche photodiode and has incorporated circuits to eliminate the range ambiguity over distances of interest. His design removes the need for calibration too.

5.5.3 *Assessment of Accuracy*

As we see in the scans above, the data is generally of excellent quality. An advantage of the AMCW sensor is that the noise characteristic is well defined, and the variance of each reading can be assessed from the amplitude of the signal at the detector. Being able to quantify the variance of an individual reading makes it possible to use an accurate sensor model. The noise distribution in ultrasound sensors is not so easy to determine. However a problem may arise when the beam is incident upon more than one surface. This problem is often called the mixed pixel problem in the vision literature. We see mixed pixel points in figure 5.12 at the large discontinuities between features. The wide beamwidth of an LED means that the points occur over quite a wide angle, but the same problem would occur (although over a narrower angle) even with a laser source. They are a serious problem in robotics applications such as free space detection.

Adams [2] has shown that although the *range* profile cannot be distinguished easily from the profile over a genuinely oblique surface, the *amplitude profile* has a distinct shape when the beam is spread over only two surfaces, as usually occurs in practice. As the discontinuity betweem surfaces is scanned, the amplitude has a well defined minimum. The second derivative is constant in value, and this value depends on the properties of the two surfaces, in particular the ratios of their distances from the sensors and of their reflectivities. These observation have allowed him to develop an adaptive algorithm to remove mixed pixel points as they are detected, and to develop the sensor to locate edges and possible targets for robot movement in a dynamic environment, as we show in Figure 5.14.

The technique provides a fast, reliable method for ranging. In a review paper, Hebert and Krotkov [90] claim that AMCW techniques are superior to any other in medium range three dimensional measurements for speed and reliability. Our work supports this view. However at lower ranges (up to about 75cm) the LEP triangulation sensor described in the next chapter, which is of comparable cost, provides the better data quality.

5.6 Summary

In this chapter we have described line of sight range sensors and the fundamental limitations to their operation. We have outlined decisions to be made in choosing an appropriate sensor technology. In line with most commercial requirements we have described the cheaper technologies in some detail with examples of current applications.

It should be clear from the chapter that no sensor can cover all circumstances. Most robots have a number of sensors; for example on the Oxford AGV we have passive vision (chapter 8), a 3-D optical range sensor (chapter 9), a 2-D optical range sensor (chapter 6) and sonar sensors (chapters 7 and 3). The development of quantitative sensor models is crucial in determining how best to combine the information from a number of sensors and is a key issue in developing sensors in the Oxford AGV project.

Chapter 6

Optical Triangulation Range Sensors for Vehicle Manœuvres

Nicholas Pears

6.1 Introduction

The close range manœuvres of an autonomous vehicle such as obstacle avoidance, parking and docking require range sensing which is robust, high bandwidth and accurate at short ranges. Range sensors based on active optical triangulation are currently among the best means of realising these requirements. Although the geometric principle of sensing is simple and proven, it is not obvious, for a given application, what sensor configuration will give the best performance. This is because there are a large number of performance trade-offs to be dealt with which interact over the boundaries of optical, mechanical and electronic sensor design.

This chapter is mainly practical in content and is divided into two main parts: The first part (sections 6.2 to 6.4) describes, in general terms, the design choices and trade-offs which must be confronted when developing an optical triangulation range sensor. The design options for optical source, image detector, and means of scene coverage are detailed in the context of the performance requirements of a given application.

The second part of the chapter (sections 6.4 to 6.8) is more specific and deals with a short range design (maximum range 2.5m) for close range vehicle manœuvres which has been implemented within the Oxford AGV project. The rangefinder acquires two dimensional data, which has been found sufficient for local maneouvres in most environments. The optical, electronic and mechanical design of the sensor is described and subsequently its performance is characterised using least squares on logarithmic plots to estimate intrinsic sensor parameters. This is necessary both to assess the success of the design and to apply meaningful algorithms to the raw range scans. Finally results are presented which demonstrate the high short range accuracy of the scanning sensor.

Further detail on sensor design, characterisation and the processing of range data can be found in Pears and Probert [158], [159] and [160] respectively.

6.2 Optical Triangulation Geometries

Scene coverage can be achieved either by scanning a spot or line stripe, or by projecting a pattern of dots or lines onto the area of interest. With the latter approach, the bandwidth of the sensor is increased and the need for potentially expensive scanning is obviated at the expense of reducing the depth of field, imposing a fixed spatial resolution on the scene and introducing a correspondence problem. Some systems have tackled this problem or the node labelling problem using coding of line thickness, space or colour. Others have utilised two or more different viewpoints of the pattern or have constructed trinocular systems consisting of one camera and two projectors [18]. The characteristics of projected pattern techniques suggest that they are most suited to fast inspection applications where there is a known and fixed depth of field over which ranging takes place. In such cases the designer can choose the thickness of the lines to prevent them breaking up and hampering the correspondence process at range discontinuities.

In scanning systems, either a light spot, or a light stripe expanded from a spot using a cylindrical lens, is scanned. Using such techniques, there is no correspondence problem, since only one portion of the scene is examined at any time. A light stripe is attractive for applications in which the scanned beam is accessible to the human eye, since it presents a much smaller eye hazard for a given amount of laser power output. Stripe projection, however, can only be used with 2-D segmented detectors such as the CCD. In addition, stripe systems have poor immunity to ambient light because the signal power is dispersed over a large image. This means that a longer integration time is required to obtain the same signal level as a spot system and, as a result, the noise due to ambient light will be increased.

The remainder of this section concentrates on geometries for scanned systems since these are more suitable for the larger depths of field required for autonomous vehicle applications.

6.2.1 Conventional Scanning Geometries

The basic geometry of scanning optical triangulation is shown in figure 6.1. It can be seen that the axis of detection is determined by object range along the projection axis, and the position of the principal point of the light collecting lens. If we assume

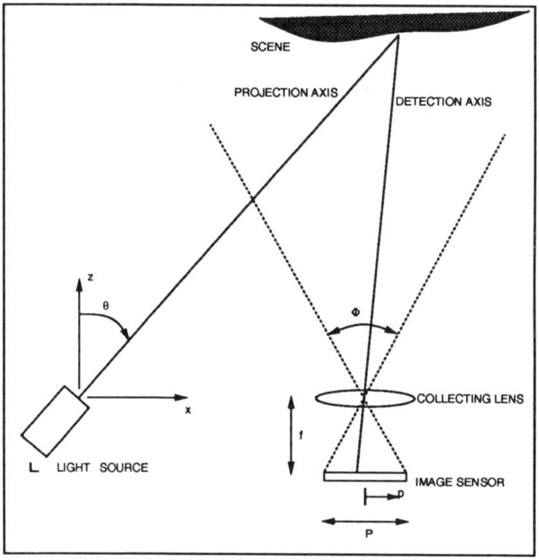

Figure 6.1: One dimensional active triangulation geometry

that the range of the object is large compared with the focal length of the collecting lens, then the focal plane is at a distance f from the principal point of the lens and simple analysis shows that:

$$z = \frac{fd}{p + f\tan\theta} \tag{6.1}$$

$$x = z\tan\theta \tag{6.2}$$

where z, p and θ are defined in Figure 6.1 and d is the x coordinate of the centre of the collecting lens. In the limit, the ratio of image resolution to range resolution is defined as the triangulation gain (G_p) and, from eq. 6.1, is given by

$$\left|\frac{\partial p}{\partial z}\right| = G_p = \frac{fd}{z^2} \tag{6.3}$$

This shows that range resolution, for a given image resolution, is proportional to source/detector separation and focal length, and decreases with the inverse square of range. In a scanned ranging system, there is an additional effect on ranging accuracy developed by the error in measurement of the projection angle. From eq. 6.1 we have

$$\left|\frac{\partial \theta}{\partial z}\right| = G_\theta = \frac{d\cos^2\theta}{z^2} \tag{6.4}$$

The effect on ranging of each of the geometrical/optical parameters of figure 6.1 can be summarised as

- *Baseline:* A small baseline gives a compact sensor and good immunity to the missing parts problem. Missing parts refers to portions of the scene in which no measurements can be made either because the projected light can not access them or because the projected light is obscured from the detector by other parts of the scene. A large baseline exacerbates the latter problem but improves range resolution.

- *Detector length and focal length:* A larger detector length can provide either a larger field of view or an improved range resolution or a limited degree of both, depending on the choice of focal length. (A short focal length gives a large field of view at the expense of accuracy and vice-versa.) However, the geometric benefit from an increased detector length must be offset with increased sensor head size and a possible degradation of electrical characteristics.

6.2.2 Synchronised Scanning Geometries

In the geometry described above, the angle between the projection axis and the optical axis of the imaging lens changes as the sensor scans across the scene. This angle defines, along with other intrinsic sensor parameters, most of the important performance measures such as minimum range, maximum range, depth of field for ranging, and accuracy.

Geometries based on synchronised scanning are employed in multi-dimensional ranging in order to maintain a more uniform triangulation geometry as the sensor scans over the scene. In addition to being a valuable property in itself, this uniformity reduces the trade-offs that have to be made in the scanning geometry of the previous section. Oomen and Verbeek [153] introduced a lateral synchronization scheme in which the optical source scanned in a direction perpendicular to the detector field of view and Rioux [177] described a longitudinal synchronisation scheme in which the scan direction is parallel to the detector field of view. The latter geometry is schematically illustrated in figure 6.2 and it can be seen that the scanned beam is tracked by an imaging mirror so that the angular separation between the projection axis and the optical axis of the imaging lens remains constant[1]. This means that the field of view of the detector rotates with the projection angle and that, to a first approximation, the scanning motion is subtracted from the image of the projected beam. Thus, the position of the image on the detector is much more dependent on

[1]Note that it is more difficult to keep the triangulation baseline constant over the scanning range for a number of reasons which include a finite imaging mirror, imaging lens separation.

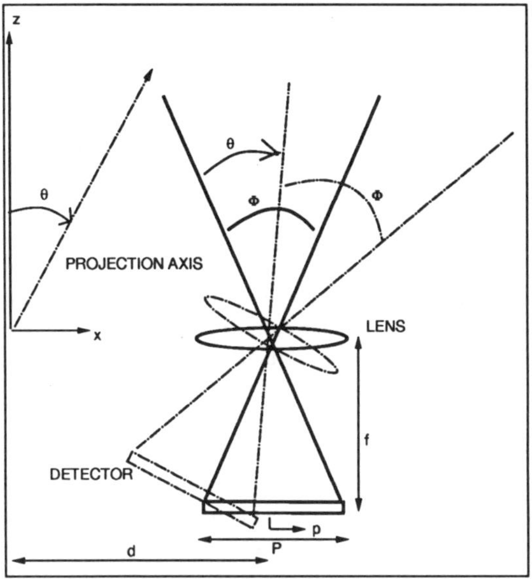

Figure 6.2: Schematic synchronised scanning geometry

range than on projection angle and nearly the whole of the detector length can be used to monitor range.

Simple geometric analysis applied to this scheme yields

$$z = \frac{fd}{p} cos^2\theta + \frac{d}{2} sin 2\theta \qquad (6.5)$$

and x is again given by eq. 6.2. The effects of image resolution and scan angle resolution on range accuracy can be described by the equations

$$G_p = \frac{fdcos^2\theta}{z^2} \qquad (6.6)$$

$$G_\theta = \frac{1}{dcos2\theta - 2z\tan\theta} \qquad (6.7)$$

Benefits of this geometry over a conventional scanning geometry include

- Reduced susceptibility to angular error in scanning (eq. 6.7) as compared to a conventional geometry (eq. 6.4).
- Focal length can be adjusted to trade off resolution and depth of field (z range), as in a conventional geometry, but now there is no reduction in field of view (x

range), since this is dependent mainly on the angle over which scanning takes place. This means that there is no limitation in the aspect ratio of the viewing area.

- Ambient light immunity is improved, because the field of view rotates with the projection angle.

6.3 Image Position Sensing Devices

In this section a comprehensive coverage of image position detection devices is not attempted. Instead the two most common devices for image position sensing within optical triangulation systems are described and compared in detail. These devices are the charge-coupled device and the lateral-effect photodiode.

6.3.1 The Charge-Coupled Device (CCD)

A charge-coupled device consists of a layer of semiconductor which is insulated from an array of electrodes by a thin layer of silicon dioxide. The operation of a CCD, for optical imaging and its many other applications, rests on principles of charge storage and charge transfer.

In optical applications, light impinging on the CCD generates electron-hole pairs in the semiconductor. The electrons are trapped in the potential well whereas the holes disappear into the substrate. The number of minority carriers (electrons for n-channel) collected under a given electrode within a given period of time (integration period) is proportional to the local light intensity. Readout of the optical image is achieved by transferring the packets of charge to adjacent CCDs which are shielded from the incident light. (If the image is read out directly from the imaging CCD, appreciable image smear can result since light is still impinging on the device during the readout period.) Clocking the charge packets out of the device is done by grouping the electrodes into three or four sets which are called phases; a different phase is applied to each set of electrodes to move the potential wells and, therefore, the charge packets across the device.

6.3.2 The Lateral-Effect Photodiode (LEP)

LEPs, also known as position sensing detectors, are less well known in the robotics community than CCD sensors.

The LEP is constructed from a slice of silicon with P and N doped layers forming a PN junction. Charge carriers are generated by light impinging on the device, are separated in the depletion region, and are distributed to the electrodes at either end of the device. Since the device has uniform resistivity in the implanted layers, photocurrent is divided according to the position of the centroid of light intensity. Thus the position of the light centroid relative to the centre of the device (see figure 6.3) is

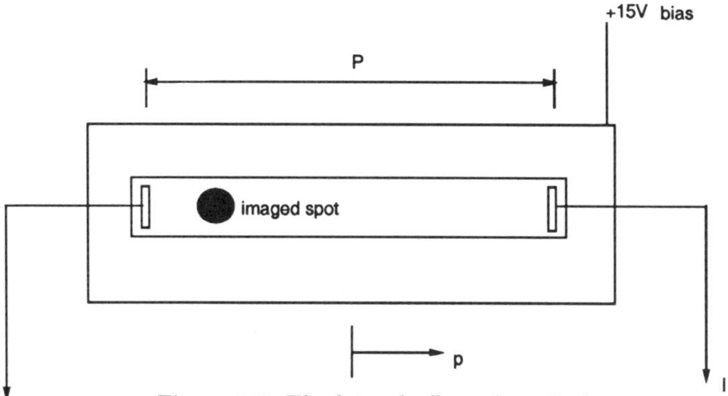

Figure 6.3: The lateral-effect photodiode

$$p = \frac{I_1 - I_2}{I_1 + I_2}(\frac{P}{2}), \qquad \{-\frac{P}{2} \leq p \leq +\frac{P}{2}\} \tag{6.8}$$

and the signal current is simply

$$I_s = I_1 + I_2 \tag{6.9}$$

6.3.3 CCD/LEP Comparison

In a CCD, image resolution is dependent on the pixel density. If pixel size is small compared to the image size, sub-pixel accuracy may be obtained by fitting a curve through a set of pixel intensity values. This, however, increases the cost and complexity of analysing the data. In contrast, an LEP is a continuum, and image position is derived from the normalised difference in current signals at the terminals of the device. The resolution of this image position is a function of the signal to noise ratio over the measurement bandwidth. This implies that at short ranges or with high power light sources, when the image irradiance is high, an LEP will outperform a CCD. At longer ranges and lower power light sources, a CCD will resolve image position more accurately. For a given application, the range must be determined above which a CCD can improve on LEP resolution in order to make a sensible choice of image sensor.

As well as image resolution, the acquisition rate for range measurements must be considered (although the two are related through temporal averaging). In a two dimensional CCD, a stripe of light is normally imaged to give two dimensional information even in a static system. Measurement is restricted to n times the frame rate, where n is the number of pixels along the length of the imaged stripe, since pixels cannot be accessed singly (typical values of 512 pixels and 25Hz frame rate gives a measurement rate of 12.8kHz for each stripe). For a one dimensional CCD, into which a spot is imaged to provide one dimensional measurements in a static system, the measurement rate is simply the pixel clock out rate divided by n giving measurement rates of a similar order of magnitude to 2-D CCDs. Again a whole row of pixels must be read for a single one dimensional measurement. In contrast the LEP makes spot measurements. The measurement bandwidth of an LEP is limited only by the capacitance and resistance of the device. For a 10x2mm 1-D LEP, this is typically around 2MHz. However in practice, measurement bandwidth in the signal processing circuitry must be restricted in the interests of signal to noise ratio.

In addition to a comparison of image position resolution, the following points must be considered when comparing the two devices

- The segmented structure of a 2-D CCD provides us with the option of scanning a stripe of light to make n range readings per frame on a profile perpendicular to the sweep direction. LEPs (1-D and 2-D), however, can only employ flying-spot coverage.

- The LEP is much more insensitive to badly collimated light sources and defocusing with range variations as compared with the CCD, since it integrates light intensity over its surface to provide image position. A CCD based system can combat these problems with peak detection circuitry and Scheimflug geometries.

- The LEP may be operated in a coherent detection scheme with a modulated source to remove ambient light and DC noise. The CCD cannot easily be used in this way and so is far less robust in applications where ambient light levels are high.

6.4 Optical Sources

6.4.1 Laser Diodes and LEDs

The coherence of the laser allows collimation into a very narrow, parallel beam of light, bringing the advantages of specific localised probing of the environment, compact scanning optics, and a small image size. Also, their spectral purity allows the use of very narrow band optical filters to reject ambient light and their very high temporal bandwidth allows high frequency square wave modulation without introducing phase shifts and signal distortion. The great problem of a laser, when projected into an open environment, is the eye safety problem, which sets an upper limit on the power output.

A light emitting diode (LED) is an incoherent light source which emits over a range of wavelengths and has a limited temporal bandwidth. Although the refractive index of glass varies with wavelength, the main problem in collimating an LED lies with its incoherence. Collimation is particularly difficult with high power LEDs since they emit over a large area. The great advantage of the LED is that it is unconditionally eye safe. This means that in open environments high power LEDs may be employed, which eliminates the signal to noise problems encountered with either noisy or insensitive detectors.

From the above discussion, it seems that lasers are preferable to LEDs in all respects except for the laser eye-safety problem which may limit power output. If high powers are required in an open environment, LEDs must be used for 1-D (unscanned systems). For scanned systems, however, higher power laser sources may be used if eye-safety is guaranteed through a laser shutdown mechanism on scanner failure and a minimum accessible distance to the laser projection.

6.4.2 Spectral Frequency of Projection: Visible, UV, or IR?

A visible beam is of great help in the alignment of any optical system and has the advantage of increased eye-safety for collimated output powers up to 1mW (class II limit) as it has the ability to stimulate the blink reflex. For beams above this power, the blink reflex is not fast enough to prevent damage to the eye. Whereas visible light is safer and more practical than infra-red, the responsivity curves of most optical sensing devices show that they are more sensitive to infra-red. Typically, an increase of sensitivity of around 50% can be obtained by moving the projected wavelength from visible light to infra-red. Infra-red is useful indoors because man made objects reflect infra-red energy well. Outdoors, however, radiation emitted by roads and other bodies limits the use of infra-red projection, particularly if the source is unmodulated. Here, UV is more useful since the ozone layer blocks sunlight which is less than 0.3 microns [64]. These factors can present a difficult trade-off since, in the interests of ambient light filtering, we need to select a light source which peaks at a spectral frequency away from that of ambient light and yet we require the detector to be responsive to that light source.

6.5 Sensor Design

The preceding sections have described the general design trade-offs that need to be confronted when building optical triangulation range sensors. This section puts these principles into practice and describes a specific implementation, designed for close range autonomous vehicle manœuvres.

6.5.1 Image Sensor and Optical Source

Since high accuracy was required at short ranges (0.5m-1m) and the maximum range was limited by the specifications embodied in the geometric design to 2.5m, the LEP was selected as the image position measurement device. The LEP is often cheaper and simpler to use than a CCD and can offer a greatly improved image position resolution at shorter ranges. The range at which a CCD system begins to outperform an LEP

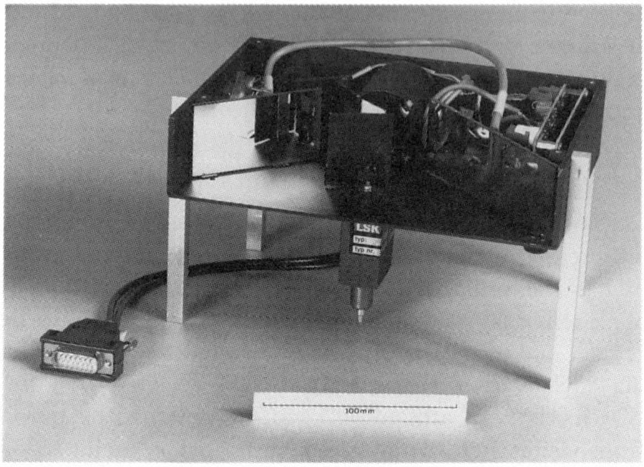

Figure 6.4: The Oxford LEP sensor

system depends on many optical, geometric, and electrical parameters. Our system requires ranges over 1m before a 1024 pixel CCD (one-pixel resolution) becomes more accurate than a simple LEP.

The optical source is a 670nm laser diode which is modulated at 10kHz and projects an average power of 0.9mW into the scene. This power complies with the class II limit (1mW) for visible lasers and is both practical and safe to work with. The sensor is shown in Figure 6.4.

6.5.2 Electronic Design

Careful circuit design for low noise performance is required for the efficient use of an LEP. The 10kHz modulation on the laser diode allows the use of synchronous detection in the receiver to maximise signal to noise ratio and eliminate DC offsets.

The key elements of the electronic design are shown in figure 6.5. The first stage, a transimpedance amplifier, is designed with three specifications in mind: low noise density (required for good image position resolution), very high gain (required to detect signals of a few nanoamps) and a wide bandwidth (required to include the first few harmonics of the 10kHz modulation). The requirement of a high bandwidth, which is a consequence of using square wave modulation in synchronous detection, is

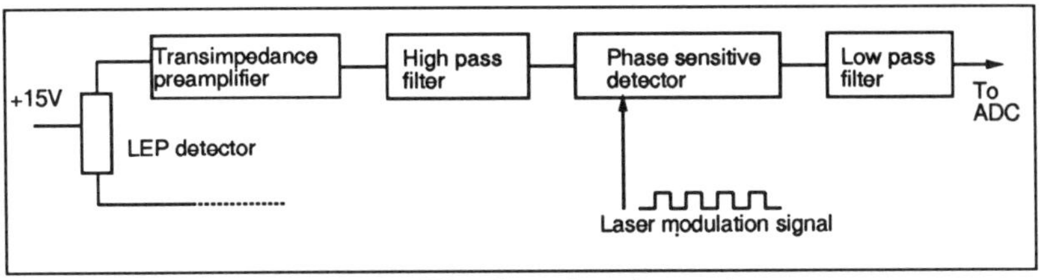

Figure 6.5: The detection electronics

in direct conflict with the need for a high transimpedance ($180M\Omega$). A typical stray capacitance of 3pf in parallel with a feedback resistor of $180M\Omega$ rolls the preamplifier response off at around 300Hz, whereas amplification of the first four harmonics of the 10kHz modulation requires a bandwidth greater than 70kHz. To allow this, the preamplification is divided into two stages: a transimpedance amplification, implemented by a ultra low noise FET op amp, with gain resistor 620kΩ, and a voltage amplification stage, implemented by a wide bandwidth FET op amp, with gain 150. A further factor of 1.9 is provided by later processing stages. The distribution of gain between the transimpedance and voltage amplification stages is such as to make their bandwidths equal. In this way, the total transimpedance is maximised for the chosen bandwidth (80kHz).

The function of the synchronous detection is to centre the signal content around the harmonics of the modulation, thus avoiding the effects of ambient light, DC offsets and drifts in the preamplification stages, and flicker ($\frac{1}{f}$) noise. This allows us to approach the theoretical limit of performance as defined by the white noise density over the measurement bandwidth. A high pass prefilter precedes the phase sensitive detector in order to allow for its non-ideal behaviour which is comprised of uneven mark-space ratio in the modulation, and imperfect matching of the two gains (inverting and non-inverting) in the phase sensitive detector.

The low pass filter shown in figure 6.5 defines the measurement bandwidth of the system. This must be set considerably below the modulation frequency, at a level which is an appropriate compromise between signal to noise ratio and speed of response. Our filter is a fourth order filter with gain 1.9 and 3db cut off at 1kHz. Since this stage dominates the dynamic response of the system, a Bessel response was chosen to avoid any ringing at edges in the range scan.

6.5.3 Geometric Design

To provide a depth of field of 0.4m to 2.5m, whilst observing the tradeoffs described in section 6.2, a baseline $d = 9.5cm$, a focal length $f = 5cm$, and a detector length

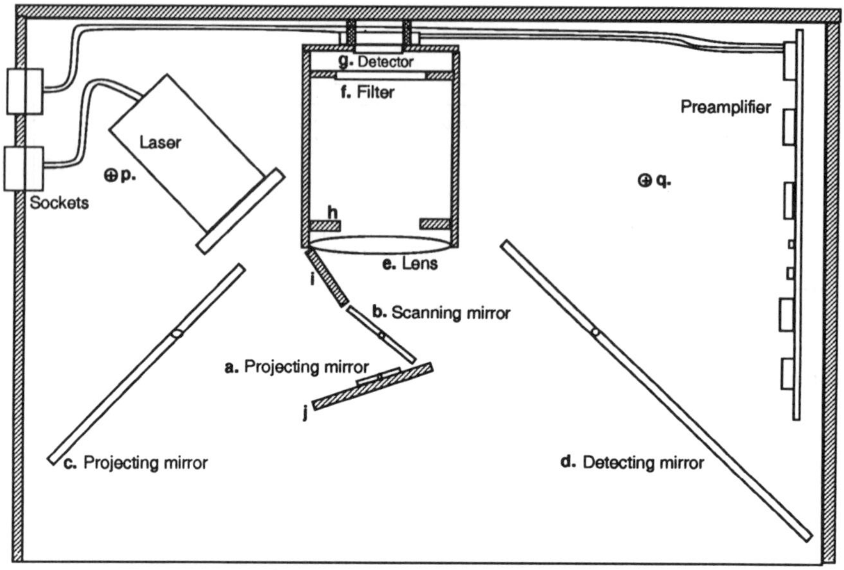

Figure 6.6: The obstacle avoidance sensor

$P = 1cm$ were used. A synchronised scanning head adapted from Livingstone and Rioux [137] was designed. Figure 6.6 shows a plan view of the sensor, which is 21cm wide, 16cm deep, and 4.2cm high. Referring to this figure, a collimated and modulated laser beam is projected onto a small mirror (a), which deflects the beam onto the front face of the scanning mirror (b). The use of this small mirror prevents the laser body from blocking the sensor aperture. The scanned beam is then projected into the scene by a larger mirror (c), which can be pivoted to set the direction of the projection angle when the scanning mirror is at rest in its zero position. With this arrangement, the laser is scanned over twice the angle over which the scanning mirror deflects, and the centre of scanning is at the virtual image point p in the sensor.

Laser light, scattered by objects in the scene, is collected by the large adjustable detection mirror (d) and is deflected onto the rear of the scanning mirror. Light leaving the scanning mirror is focussed by the lens (e) and passes through an optical filter (f), matched to the laser wavelength, before forming an image of the projected spot on the lateral-effect photodiode (g). To minimise noise, the detector signals are amplified inside the camera head before being passed to the synchronous detector in the sensor interface rack.

With the geometry described above, the lens is effectively scanned around virtual image point q in the sensor on an arc with radius equal to the separation between the scanning mirror and the lens. In our design, this separation is kept as small as

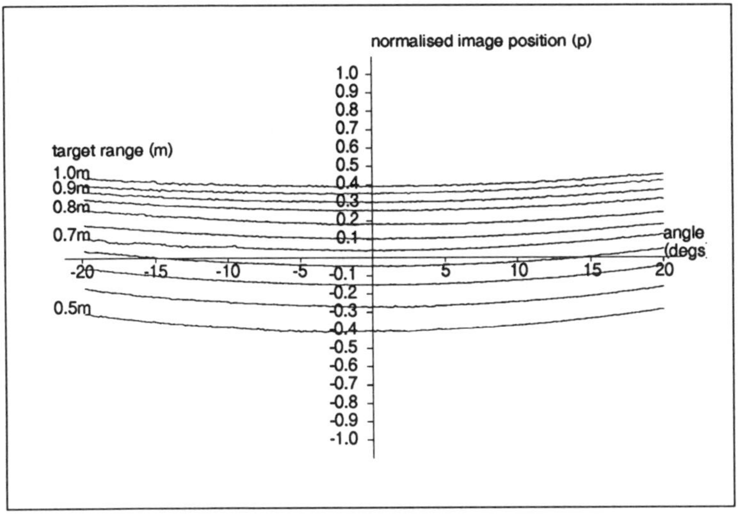

Figure 6.7: Calibration curves: image position against scan angle

possible to minimise variations in triangulation baseline over the scanning range. The dimensions and positioning of the detection mirror (d) are critical and ensure that the full sensor aperture (the full surface of the scanning mirror) is accessible over all combinations of scan angle and target range.

The sensor is designed with as large an aperture as possible that is consistent with our scanning requirements because of the LEP's dependence on a good signal to noise ratio. Most of the aperture derives from the depth of the scanning mirror (4cm) rather than its width (2.2cm) in order to limit rotational inertia. Direct optical paths, in which laser light is focussed directly from the scene onto the LEP, have been prevented with the use of a 'cats-eye' aperture stop (h) behind the lens and shielding plates (i) and (j).

Figure 6.7 shows the recorded image position as the sensor scanned across a planar target at ranges of 0.5 to 1.0m. Since the lens is scanned in synchronism with the laser, there is only a small variation in image position over the scanning range. The fall in triangulation gain with target range is shown by the decreasing separation of the image position curves.

The final specifications of the sensor are: field of view (scan range) 40 degrees, depth of field (z range) 0.4m to 2.5m, samples per horizontal scan 256, scan frequency 10Hz, sample frequency 2.56kHz, projected laser power 0.9mW at 670nm (class II), detector bandwidth 1kHz (set by fourth order Bessel filter).

6.6 Sensor Characterisation

A comparison of the actual performance of the sensor and the performance predicted from the specifications of the LEP, preamplification electronics and sensor geometry is essential to determine how close the sensor is to the physical limits of ranging accuracy. In the following analysis, only the effect of image position resolution on ranging resolution is considered. However, in a scanning sensor, errors in scan angle measurement will also contribute to ranging error. The specifications of our scanner quote a mirror angle repeatability of 0.1mrad which makes that the effect of scan angle resolution on accuracy negligible.

6.6.1 *Experimental Conditions*

The sensor was set up with the laser pointing along the z-axis and image position, p, was calibrated against range, z. A target, consisting of an off-white cardboard box was set up at ranges between 0.75m and 2.5m in steps of 0.25m. At each range, 1000 measurements of signal current, I_s, and normalised image position, p, were made. For each image position measurement, range, z, was interpolated from the calibration look up table.

6.6.2 *System Noise Estimation*

Using logarithmic plots of averaged signal current against range and image resolution (the standard deviation of image position) against range, the effective noise density in the system can be estimated.

If it is assumed that the projected spot is small enough and distant enough to be treated as a point source and that purely Lambertian scattering occurs in the scene, then the total signal current I_s at range z, for laser power P_l, can be approximated by

$$I_s = \frac{S}{z^2} \tag{6.10}$$

The signal current constant, S, is given by

$$S = \frac{\rho A \cos\theta_d T_1 T_2 R_\phi P_l}{2\pi} \tag{6.11}$$

where ρ is scene reflectivity, A is aperture (m^2), θ_d is the orientation of the aperture with respect to the target surface, T_1 is the reflectivity of the projection optics, T_2 is the reflectivity of the detection optics, R_ϕ is the responsivity of the LEP (A/W) and P_l is the power projected by the laser (W).

Rather than measuring or estimating these individual parameters, in which case the errors in each value would contribute to the error in the estimation of the sensor performance, the composite parameter, S, can be estimated accurately from a logarithmic plot of signal current against range. This plot is given in figure 6.8 where

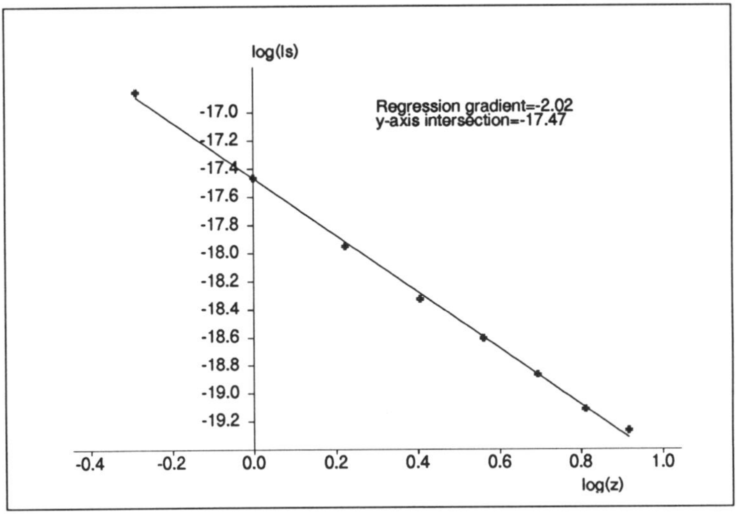

Figure 6.8: Logarithmic plot of signal current against range

the crosses represent the actual measurements and the solid line is a least squares fit. The intersection of the regression at the y-axis gives

$$S = 2.59 \times 10^{-8} Am^2 \tag{6.12}$$

Now, image resolution for an LEP is

$$\Delta p = \frac{P}{2\frac{I_s}{I_n}} \tag{6.13}$$

where P is the detector length and $\frac{I_s}{I_n}$ is the signal current to noise current ratio of the detection. Since image position measurements are normalised to lie in the region $\{-1 \leq p \leq 1\}$, then

$$\Delta p_{norm} = \frac{I_n}{I_s} = \frac{I_n z^2}{S} \tag{6.14}$$

where I_n is the total noise current over the measurement bandwidth. Thus a logarithmic plot of image resolution against range, given in figure 6.9, gives the constant $\frac{I_n}{S}$ and thus the noise current I_n

$$I_n = 36.5pA \tag{6.15}$$

In these tests, the measurement bandwidth was set by a fourth order low pass Bessel filter with cut off at 1kHz. Thus the approximate current noise density, assuming an ideal filter response, is

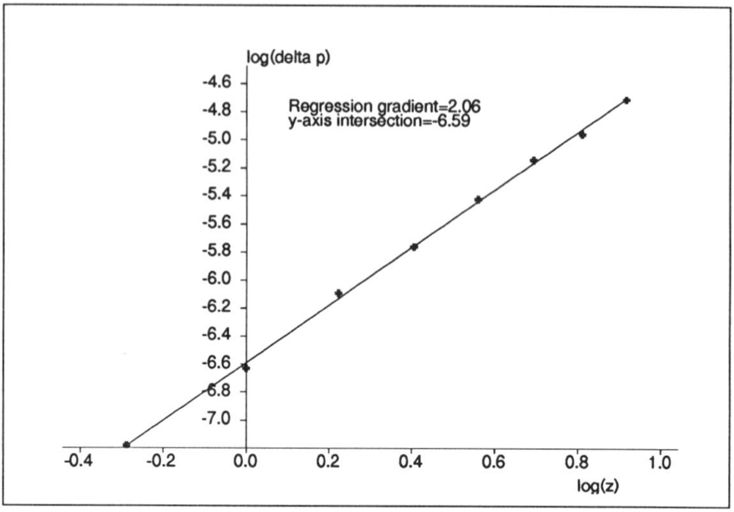

Figure 6.9: Logarithmic plot of image resolution against range

$$i_n = 1.15pAHz^{-\frac{1}{2}} \tag{6.16}$$

6.6.3 Theoretical Noise Calculation

In calculating the theoretical noise density, it is assumed that shot noise due to the signal current, and shot noise due to any current generated by ambient light passing through the optical filter, are negligible.

The total detector noise density is the RMS summation of shot noise from the dark current and thermal noise. For the LEP used in the sensor, dark current $I_d = 100nA$, $R_s = 50k\Omega$, and assuming $T = 298K$:

$$i_{nd} = \sqrt{2eI_d + \frac{4kT}{R_s}} = 0.6pAHz^{-\frac{1}{2}} \tag{6.17}$$

Preamplifier noise, including feedback resistor noise is given as:

$$i_{na} = \sqrt{i_{na}^2 + \frac{e_{na}^2}{R_s} + \frac{4kT}{R_f}} = 0.2pAHz^{-\frac{1}{2}} \tag{6.18}$$

Calculating the RMS of eq. 6.17 and eq. 6.18 gives the total current noise density (I_n) for the detector-preamplifier combination as $0.632pAHz^{-\frac{1}{2}}$. The noise density calculated from measurements ($1.15pAHz^{-\frac{1}{2}}$) is only 82% above this value which shows

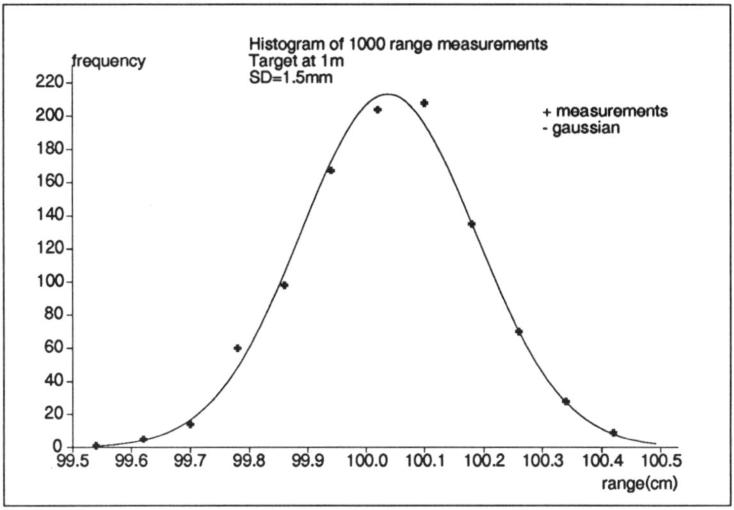

Figure 6.10: The distribution of range measurements at 1m

that the use of synchronous detection has given a reasonably close agreement between the theoretical performance limit of the sensor and its actual ranging performance.

6.7 Sensor Performance

6.7.1 *Measured Accuracy*

Figure 6.10 shows a histogram of range measurements when a target was placed at 1m from the sensor, and 1000 readings were taken. In this figure, the crosses show the frequency with which a measurement fell within a particular range interval. The solid line shows the equivalent Gaussian, generated using mean and standard deviation of the batch of 1000 measurements. From figure 6.10 it is evident that, in any processing of the raw sensor data, the assumption of of a Gaussian form is reasonable. Standard deviations for all target ranges, and their values as a percentage of the target range, are shown in Table 6.1.

6.7.2 *Predicted Accuracy*

In order to calculate range resolution from image position resolution the triangulation gain constant, fd, must be estimated. Combining eq. 6.6 and eq. 6.14) for $\cos\theta = 1$ gives

$$\Delta z = T z^4 \tag{6.19}$$

Range(m)	Standard deviation (cm)	SD as percentage of range
0.75	0.05	0.067
1.0	0.148	0.148
1.25	0.463	0.37
1.5	0.798	0.532
1.75	1.568	1.1
2.0	2.547	1.27
2.25	3.713	1.65
2.5	7.662	3.06

Table 6.1: Ranging Accuracy

where

$$T = \frac{PI_n}{2fdS} \tag{6.20}$$

Thus, a logarithmic plot of Δz against z will give the constant T and, since I_n and S are known from previous logarithmic plots, fd can be obtained. The intersection of the regression line with the y-axis in such a plot gives $fd = 0.0044m$.

Range resolution against range can now be plotted for the measured noise density ($1.15\ pAHz^{-\frac{1}{2}}$) and the theoretical noise density ($0.632\ pAHz^{-\frac{1}{2}}$) using eq. 6.19 and eq. 6.20. These two plots, and the points at which ranging accuracy was measured (from table 1), are shown in figure 6.11.

When scanning, ranging accuracy will be poorer at the edges of the scan than at the centre since, from eq. 6.6, triangulation gain drops with $\cos^2\theta$ and, to compound this, signal strength will be reduced because of the apparent reduction in aperture with $\cos\theta$. The total effect is to reduce ranging accuracy at the edges of a scan to around 83% of its value at the centre. A typical plot in figure 6.12, shows the excellent response over abrupt edges and indicates that the variation in ranging accuracy with scan angle is small compared to the variation with range.

6.8 Use of the Sensor within the Oxford AGV Project

The sensor is being used within the Oxford AGV project to investigate

- How intensity data can be used in conjunction with range data in order to identify and locate range features.
- How active sensing strategies can be used to facilitate robust close range autonomous vehicle manœuvres.

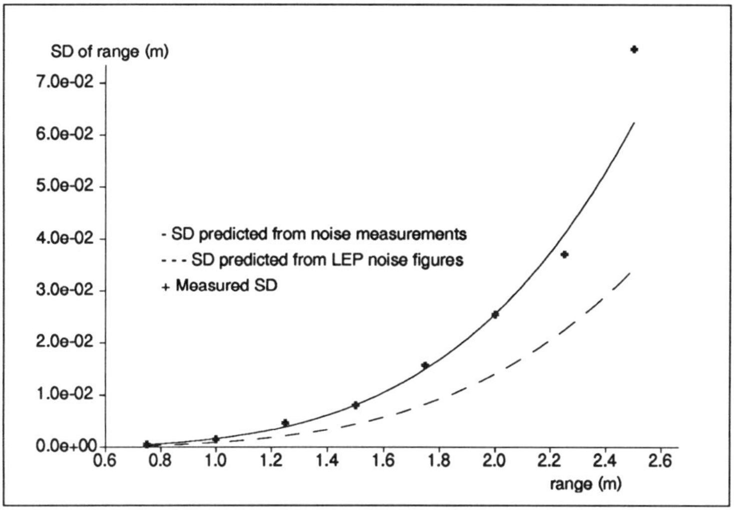

Figure 6.11: Plot of predicted and actual SD of range against range

6.8.1 The Use of Intensity Data

An attractive feature of the LEP is the Gaussian noise characteristic of image position measurement and the direct relationship between its variance and the intensity signal. These facts make the sensor particularly amenable to statistically based feature detection, tracking and data fusion algorithms such as the Kalman filter.

Experimentation with the sensor has shown that intensity data may be used to detect certain qualities of, or classes of, feature. These include

- Features from intensity signatures: It has been noted that signal amplitude data may be useful for identification and localisation of certain types of feature. Figure 6.13b illustrates that, in the case of a cylindrical object, a distinct peak is found at the centre of the feature. This may be understood in terms of Lambert's cosine law for diffuse reflection.

- Range discontinuities: Intensity data can be very useful in detecting small range discontinuities since it is zero when the projected spot is occluded by the discontinuity (missing parts). More generally, intensity can be used to form an accurate validation gate for range discontinuity detection within the Kalman filter algorithm.

- Feature reflectivity: If the sensor is calibrated with a target of known reflectivity, feature reflectivity can be estimated from intensity and range data.

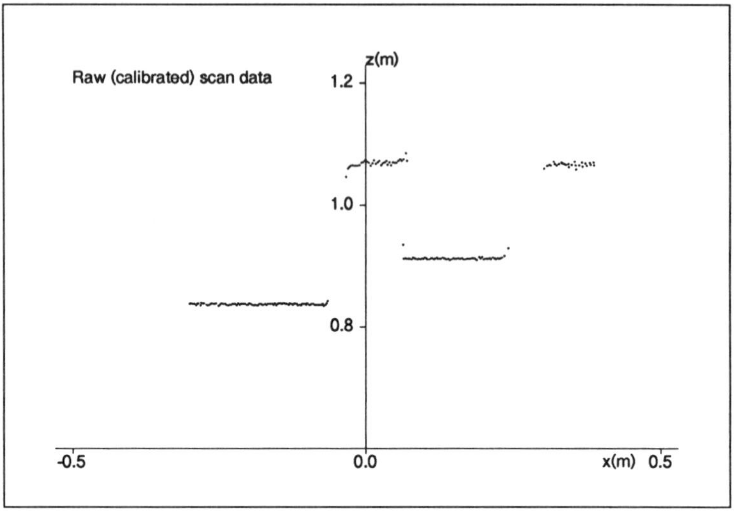

Figure 6.12: Range image as the sensor scans over a scene incorporating several cardboard boxes.

6.8.2 Active Sensing

The full implementation of the sensor described in this chapter is active, both in the sense that it uses structured light and in the sense that it can actively look for and track features, which are predicted to be observable, rather than waiting for them to enter a static field of view. When operating actively, it can rotate its whole body in a horizontal plane between +90 and -90 degrees and can vary the scan of the laser over any angular range between -20 and +20 degrees.

The sensor has been developed as a locally autonomous sensor agent (see chapter 3) in which the software for calibration, filtering, feature extraction, feature tracking, sensor planning and head control are mounted on a small network of transputers dedicated to the sensor. The local intelligence of the sensor is embodied in the local sensor planner which receives a state vector describing the vehicle motion over a transputer link connected to the vehicle controller. Using this information and the sensor's own information concerning features and their movements, it can decide which feature tracks to maintain based on the costs and utility of active sensing actions. A vector describing the local environment is passed back from the sensor network to the vehicle's path planner.

This active realisation is being used to demonstrate close range manœuvres such as obstacle avoidance on a Robosoft mobile platform.

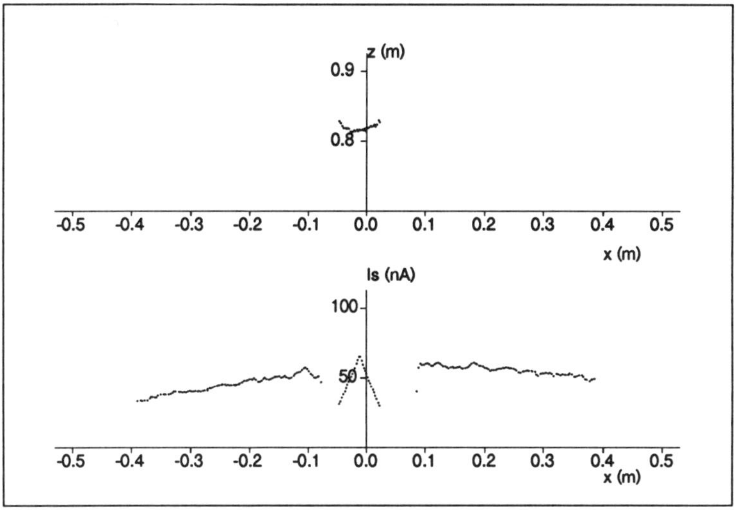

Figure 6.13: Range image (top) and intensity image (bottom) as the sensor scans over a scene incorporating a cylinder.

6.9 Conclusions

Optical triangulation range sensor design has been discussed with a view to implementing a short range sensor for close range autonomous vehicle manœuvres such as obstacle avoidance, parking and docking. This discussion has shown the implications of choosing a certain sensor configuration (optical source, image sensor and means scene coverage) in terms of performance criteria such as accuracy, bandwidth and useful range.

A wide field of view range sensor for short range mobile robots manœuvres has been described. Measurements have shown that ranging is accurate to 0.15% at 1m, 1.3% at 2m, and 3% at 2.5m for a measurement bandwidth of 1kHz, which is suitable for real-time planning, and an average projected power of 0.9mW, which is eye-safe. These performance figures compare well with other rangefinders available at comparable cost. In addition, we have extended the application range of LEPs, which are normally used with non eye-safe lasers over much smaller depths of field.

A sensor characterisation showed that the measured range resolution is only 82% above the theoretical performance limit as defined by the detector and preamplifier noise specifications. This good performance is the result of careful preamplifier design and the use of synchronous detection. Also, the sensor characterisation allows a good estimate of range variance to be made by measuring the received photocurrent. This variance estimation, along with an approximate Gaussian form on the ranging errors, makes the sensor particularly amenable to statistically based feature detection, tracking and data fusion algorithms.

Chapter 7

Modular Sonar Sensing for Vehicle Navigation

James Manyika

7.1 Introduction

In current research on autonomous robot vehicles, a variety of sensor technologies are used, ranging from infra-red devices, laser range-finders, sonar, CCD cameras and gyroscopes. Sonar, in the form of the standard Polaroid device, is the most widely used sensor in mobile robotics. While advanced research has produced sophisticated sonar sensors capable of complete imaging, the standard Polaroid due to its extremely low cost and ease of use, has remained a popular choice in mobile robotics research. Typical uses of sonar include, obstacle detection and avoidance [100][24], and navigation and map-building [56][133]. The application of sonar to these tasks poses a number of problems. In particular, it is well known that range returns from sonar sensors are subject to distorting effects caused by long wave-lengths and wide beam-widths, making it difficult to apply any "ray-tracing" type algorithms to the interpretation of sonar data. Furthermore, because the rate at which range information can be obtained is physically limited by the speed of sound in air, a vehicle using sonar sensors to obtain map or guidance information is typically forced to adopt a "stop-look-move" motion strategy during task execution.

Therefore the effective use of sonar depends, firstly, on a proper interpretation of the measurement data, secondly, on the development of a way of overcoming the limitations of sonar, and thirdly on the improvement of sonar data processing. This chapter addresses precisely these issues and presents a model for a novel sonar sensor which overcomes some of the limitations of conventional sonar devices. The result is the realization of a *Tracking Sonar* with the ability to focus attention on environmental features in real-time. This leads to the new ways of implementing mobile robot localization and map-building which we describe.

7.2 Sonar Model

7.2.1 Region of Constant Depth Model

In developing our model we exploit the observation made by Kuc and Siegel [124] that at the wave-lengths typical of in-air sonar, common indoor environments consist predominantly of "mirror-like" reflectors which correspond to features in the environment[1]. This results in range-bearing scans composed of regions in which the measured range is constant. It has already been well-documented that given knowledge of these *Regions of Constant Depth* (RCD), accurate (sub-centimetre) and reliable maps of indoor environments may be constructed from sonar data [125][133]. An RCD is "generated" in the following way; when a scanning sonar device is aimed perpendicular to an environment feature, the range returned is indeed the true range to that feature. However, as the sonar is rotated away from the perpendicular, the center of the main lobe of the sonar beam is reflected off the feature in such a way that it is not returned to the receiver. Instead, a reflection from that part of the main lobe which is now perpendicular to the feature is returned to the receiver. Thus, the same range measurement is recorded as the sonar is rotated through an angle equal to the main-lobe beam-width. The result of this is that a complete range-bearing scan is composed of a set of RCDs, one associated with each environment feature. Figure 7.1 shows a typical sonar scan together with a map of the room from which the scan was taken. A semi-circular RCD associated with each main environment feature, wall, corner or edge, can be seen.

The precise width of the RCD is determined by various physical parameters of the propagated beam; decreasing the threshold increases RCD width, decreasing the frequency of the transmitted beam increases the RCD width. This can be illustrated by impulse response analysis as shown in [124][139]. Successive range returns constitute an RCD if the absolute range difference between the minimum and the maximum of the connected set of ranges is less than some ϵ_r. The error ϵ_r is typically of the order of the precision with which range can be measured. The range of the RCD r_{RCD}, is given by the mode of the distribution of the ranges making up the RCD. We can define β_{RCD} the width of the RCD and β_{min} the minimum acceptable width for what constitutes an RCD. This RCD model is illustrated in Figure 7.1(b) for a

[1] These "mirror-like" reflectors have been called specularities in the literature [24][132]. However, strictly speaking, specularity results from the highly differing acoustic impedances of air and solids [147] whereas the mirror-like reflectors are due to the wavelengths of in-air sonar and the wide beam-widths coupled with the thresholding nature of the Polaroid circuit.

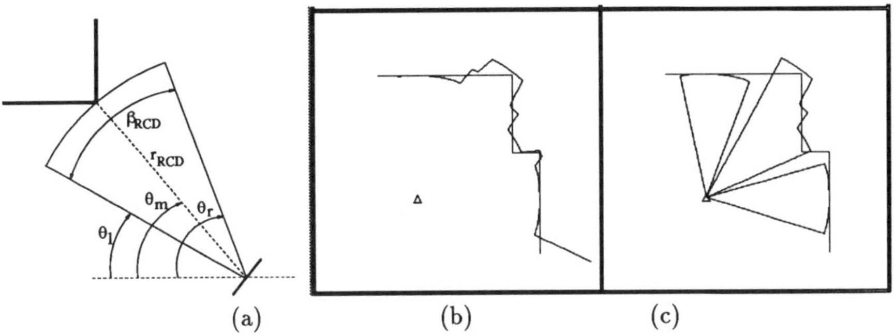

Figure 7.1: *(a) Parameters in RCD model. Sector scan showing (b) range information (c) RCDs extracted*

symmetrical beam[2]. RCDs may be extracted from a range-bearing scan by simply segmenting range returns sequentially according to the criteria ϵ_r and rejecting those whose width is less than β_{min}.

Despite the better understanding provided by the RCD model, we can identify the following shortcomings of sonar:

1. **Imprecise target bearing.** In the absence of a scan which establishes the angular extent of the RCD, the precise bearing of the target is not known. This is exacerbated by the instability of RCD edges.

2. **Characterisation of targets.** The range and edge-angles defining an RCD $(\theta_l, \theta_r, r_{RCD})$, give no indication as to the geometric nature of the target.

3. **Sonar data processing.** Since an RCD may be characterized by a single range value together with the bearing of its centre (See Figure 7.1(b)), the vast majority of sonar range readings do not provide any useful additional information about the feature causing a reflection.

7.2.2 RCD-Monopulse Sonar

The sensor that we develop to overcome these limitations makes use of two transducers slightly displaced and measuring the same returned signal. The use of two antennae in this way is employed extensively in tracking radar systems [185][17], and the difference signal measured is commonly called the *monopulse* signal. In radar the difference signal is either constructed from a comparison of the returned signal amplitudes (called amplitude-mono-pulsing) or from a comparison of the returned signal phase (called phase-mono-pulsing) [17]. A similar approach can be used with sonar assuming

[2]In practise the Polaroid beam is not symmetric and varies from device to device.

that we have devices which are able to measure amplitude and phase. This is not possible with the standard Polaroid device but can be implemented using modified sonar devices as shown by Barshan and Kuc [13].

While not true phase-monopulse, our approach is analogous to the phase-monopulse technique in providing a measurement of the difference in time of arrival of a common signal at two spaced receivers. For this reason we refer to the sonar device as an RCD-monopulse sensor or more accurately *Differential Sonar*. The differential sonar consists of two standard Polaroid transducers mounted on a common baseline on a high-speed servo as shown in Figure 7.2. Figure 7.3 illustrates the principle. One of the transducers (T/R) both transmits and receives acoustic signals, the second transducer (R) acts only as a receiver. Therefore, the following is available; (i) a conventional time-of-flight range measurement and (ii) a measurement of the difference Δ in time of arrival between the two receivers.

When the target feature is centered between the receiver pair, the difference in signal arrival times is zero. As the target moves away from center, the difference signal grows linearly with off-axis bearing. The exact bearing to the target feature can be determined by rotating the sonar pair so as to null the time difference signal and simply reading off the resulting aim direction. If the feature moves, or more commonly in mobile robotics, the platform on which the sonar is mounted moves, the same signal can be used to maintain track on the target feature as we shall show.

7.2.3 Differential Feature Model

The differential time of flight, Δ, is obtained by measuring the time between the reception on the first device to receive, *primary reception*, and the time when the second device receives, *secondary reception.*

Given that r is the range to the feature, d the separation of the T/R and the R devices and α the angle of inclination from the perpendicular to the plane of the feature, we can define the differential for angle α within the angular extent of the RCD. For the plane shown in Figure 7.3, the differential can, from geometrical considerations, be shown to be given by

$$\Delta_p = \frac{1}{c}\left[\sqrt{r^2 + \frac{d^2}{4} - rd\ \sin(\alpha)} + \sqrt{r^2 + \frac{d^2}{4} + rd\ \sin(\alpha)} - 2\left(r - \frac{d}{2}\ \sin(\alpha)\right)\right]. \tag{7.1}$$

The form of the differential Δ can also be derived for the other basic feature types naturally occurring in indoor environments; corners, edges and walls [139]. For these features, the theoretical Δ is linear within the width of the RCD. However, there are slight differences arising from the geometry of the reflecting surfaces and the asymmetry of the transducers. These differences are discussed in [139] but given our hardware these are not noticeable and can be ignored. Figure 7.4 illustrates an RCD extracted for a plane feature and the associated differential.

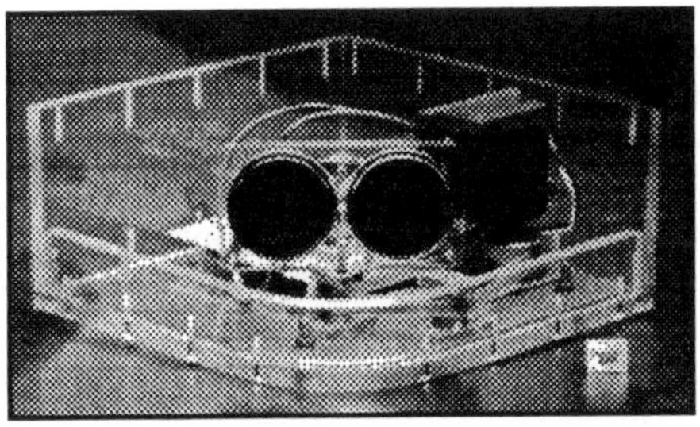

Figure 7.2: *Prototype RCD-Monopulse unit*

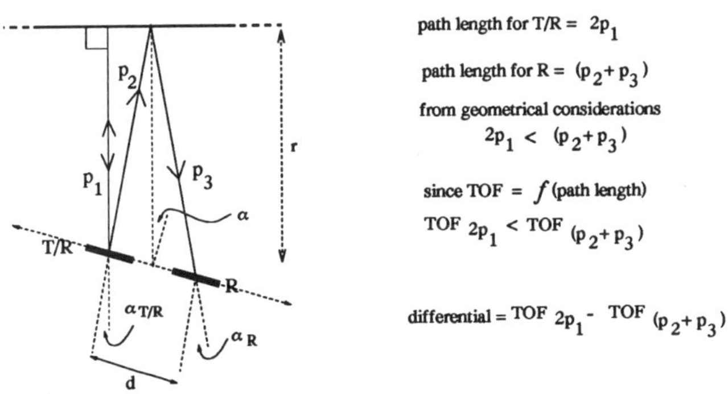

Figure 7.3: *Illustrating the differential principle for a positive value of* α

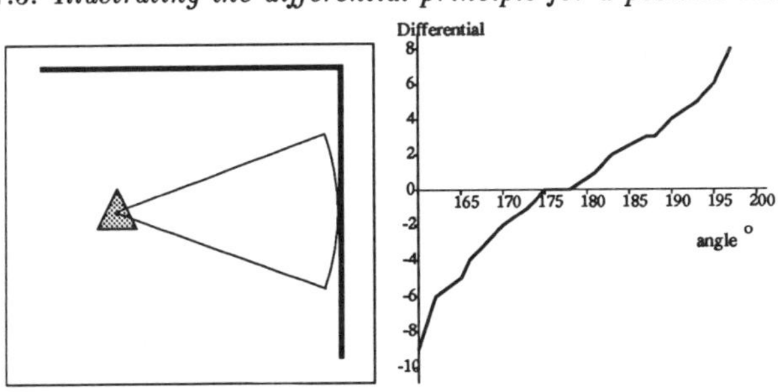

Figure 7.4: *An extracted plane RCD and the associated differential. (a) shows the RCD and (b) the differential. The vertical scale in (b) is scaled by* $3.2\mu s$ *and the horizontal axis is the angular extent of the RCD in degrees*

7.3 A Tracking Sonar

Each return from the differential sonar provides the following: $[r,\ \theta,\ |\ \Delta\ |,\ sign(\Delta)]$. Given that the return is within the bounds of an RCD, i.e. $\theta_l \leq \theta \leq \theta_r$, and assuming an ideal linear differential, θ is a computable angular distance away from the precise location of the target θ_m (See Figure 7.1(a)) in a direction indicated by $sign(\Delta)$. Therefore, a simple tracking algorithm can be implemented which continually "corrects" for each non-zero differential by moving towards the zero differential.

7.3.1 Differential Uncertainty and Real-time Tracking

In typical application environments, it turns out that the measured differential is not as linear as predicted and as shown in the controlled measurement of Figure 7.4. This is illustrated in the results of Figure 7.5 taken in a typical cluttered environment.

To further highlight this uncertainty, it is observed that if the differential sonar is fired repeatedly at the same orientation within the bounds of the RCD the differential measured is not constant as would be expected. This is illustrated in Figure 7.6 where several hundred observations are taken.

From the differential plots shown in Figures 7.5 and 7.6(a), it can be seen that some smoothing or filtering operation is required to remove outliers and spurious readings given that we can predict the form of the differential characteristic. We treat the measured differentials as noise corrupted observations of the true differentials and use a Kalman filter to estimate the true differential [139]. We make the assumption that the noise is zero mean and uncorrelated. The results in Figure 7.6(b) show the measured and estimated differentials.

Using the differential estimate described above, we can now implement a real-time Tracking Sonar. The servo provides a control input which has the effect of reducing the differential by steering towards the centre of the RCD. We can define $s(k)$ as the servo step at time k in correcting for the differential. The servo step can be based on the angular distance from the true orientation, however better performance is obtained by correcting by a smaller step due to hardware considerations. We introduce a parameter ψ which defines a ***dead-band*** in which the servo does not correct for the differential. The Tracking Sonar algorithm is as shown in Figure 7.7. The criteria for RCD selection depends on the application or goals of the system and shall be discussed later. The results in Figure 7.8 show the performance of the tracking algorithm for a target which moves and stops as indicated.

7.3.2 Probabilistic Model and Error Analysis

The Tracking Sonar provides a measurement vector $\mathbf{z} = [r, \theta]^T$, where the orientation θ is referenced to a *datum* line on the sensor itself. We assume the uncertainty in the measurements is in the form of additive noise. Therefore, we can write that

$$\mathbf{z} = \mathbf{z}_{true} + \mathbf{v}(\zeta, \mathbf{z}_{true}), \tag{7.2}$$

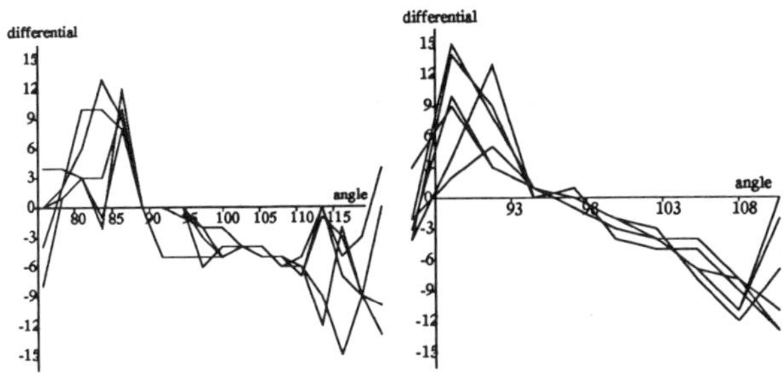

(a) Plane (b) Edge

Figure 7.5: *Plots showing multiple (5) differentials within the width of the RCD for a plane and an edge. The angle axis is the orientation from some fixed datum of the ranges making up the RCD. Only differential information within the RCD width is of significance. The vertical scale is the differential TOF* $\times 3.2\mu s$

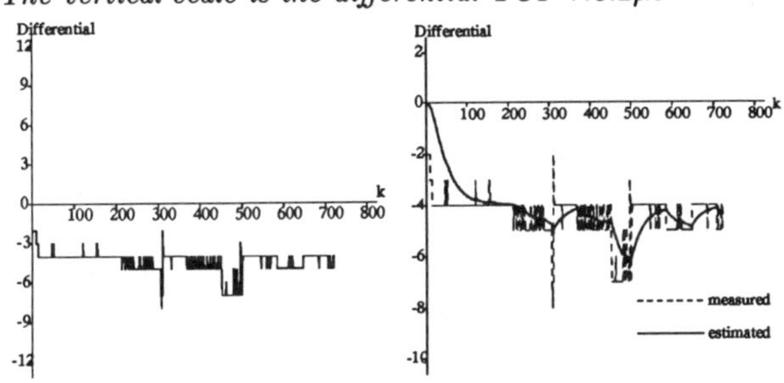

(a) (b)

Figure 7.6: *(a)The differential at a negative offset within the bounds of an RCD for a given feature. (b) The measured and estimated differential at a negative offset. All values of the differential being* $\times 3.2\mu s$

where $\mathbf{z}_{true}$ is the true state being measured and $\mathbf{v}$ the observation noise which is dependent on the true value and on ζ the sensor control parameters (differential filter parameters, dead-band *et cetera*). It is therefore desirable to develop a model of the additive noise $\mathbf{v}$ for a given range of sensor control parameters. Such a model can be developed by considering the factors determining the performance and uncertainty in the Tracking Sonar observations and making use of spectral estimation techniques. The factors determining performance are

- Differential filter parameters. The observation noise model for the measurement of the differential, the transition model based on the correctional motion of the servo, etc.

- Differential dead-band and other hardware constraints.

In modelling data from the Tracking Sonar, the range r and orientation θ over time can be treated as random processes. If there is no relative motion between the feature being tracked and the sensor, then the random processes can be viewed as stationary[3]. In modelling information from the sensor we shall only consider the stationary case[4]. Our interest in the proceeding analysis lies only in determining whether the process noise is white or not. Random processes are defined in terms of their ensemble averages which can be estimated [154]. Our model shall be in terms of such averages as estimated using finite sequences.

As an illustration, we consider the statistical averages when a particular operational parameter, the differential measurement noise variance σ_δ^2, is varied. Figure 7.9 shows variations in θ with observation noise variance σ_δ^2. The results show that for a stationary feature the variance in the orientation of the target is reduced with a larger observation noise variance σ_δ^2. Increasing the noise variance eventually results in some instability [141]. A compromise has to be reached between increasing the noise variance and having a small enough variance to give a good response to motion.

To determine the statistical model of the uncertainty $\mathbf{v}$, we consider the autocorrelation and power spectra of the data streams for various operational parameters. Figure 7.10 shows the autocorrelation and power spectrum for a given value of the differential observation noise variance σ_δ^2 and the transition model σ_u^2. The results in Figure 7.10, based on the confidence tests described by Jenkins and Watts [110] and the discussion in [141], justify a white noise model for measurement uncertainty $\mathbf{v}$. In a similar manner, the modelling of sensor uncertainty as white can be shown to hold for values of operational parameters giving satisfactory tracking performance.

[3]This is not strictly true, because the differential used to maintain track, is essentially the result of a range measurement. Closely timed range measurements in the short interval, are not really stationary due to the effect of temperature. However, the error due to this is small enough to be ignored in our case.

[4]The case when there is motion requires that we consider the residuals of the process given an adequate process model, and attempt to describe these in terms of their averages by considering the residuals as a stationary process [110].

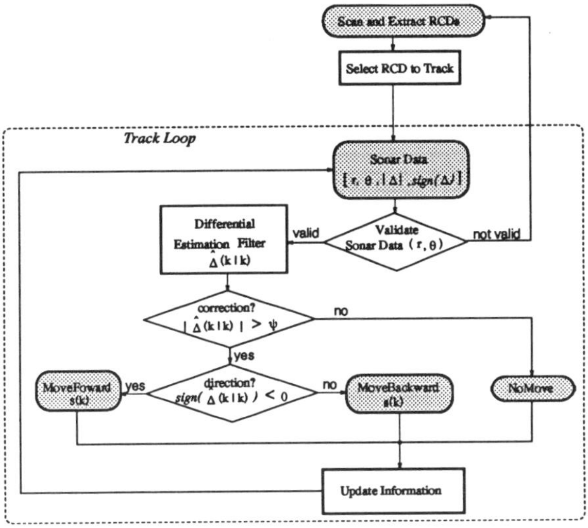

Figure 7.7: *Tracking Sonar Algorithm. The shaded areas indicate hardware related procedures. The selection of RCD to track is application dependent and shall be discussed in the next Chapter*

Therefore, the noise vector $\mathbf{v}$ (from Equation 7.2) has a 2-by-2 covariance matrix $\mathbf{R}$ composed of the variances σ_r^2 and σ_θ^2 along the diagonal. We assume that noise associated with the range measurement v_r is uncorrelated with the noise associated with the orientation v_θ. This assumption is in the absence of a modelling of the possible cross-correlations between v_r and v_θ at the same timestep. The variances in the matrix $\mathbf{R}$ also depend on the measured quantities themselves and hence the matrix is more accurately written $\mathbf{R}[k, \mathbf{z}(k)]$. Such an observation dependent covariance $\mathbf{R}[k, \mathbf{z}(k)]$ is sometimes known as a *location covariance.* A model of the variation of the variances with the measured quantities can be developed by analysing the variances for a variety of measurements within the operating range.

7.3.3 Performance and Limitations

The real-time performance of the tracking sonar in fixating on a given feature depends on several factors:

Speed of relative motion. The limited sonar return rate coupled with the fact that the servo requires a finite time to effect correctional motion, places a limit on the data rate of the sensor. The maximum data rate of the sensor is of the order of $30Hz$. Relative motion of a feature with regard to the sensor, can be decomposed into the angular motion and the linear motion along the normal to the feature. Assuming

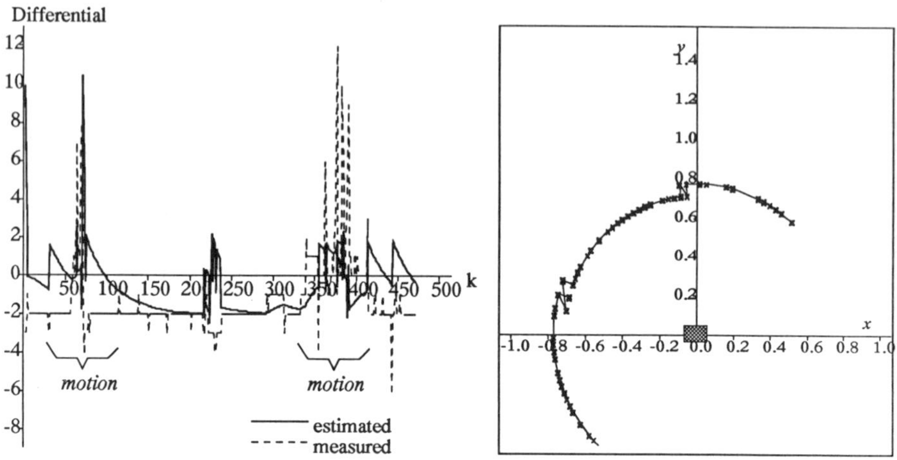

(a) (b)

Figure 7.8: *(a) Results showing the measured and estimated differentials in the tracking of an RCD target moving at arbitrary time steps (50-75 and 350-390) with the differential sonar platform stationary. There is a no-correction differential dead-band of 2 (×3.2 μs). (b) Illustrating rotational motion of a sensor in ego-centric coordinates as calculated from the range and orientation while tracking a stationary feature*

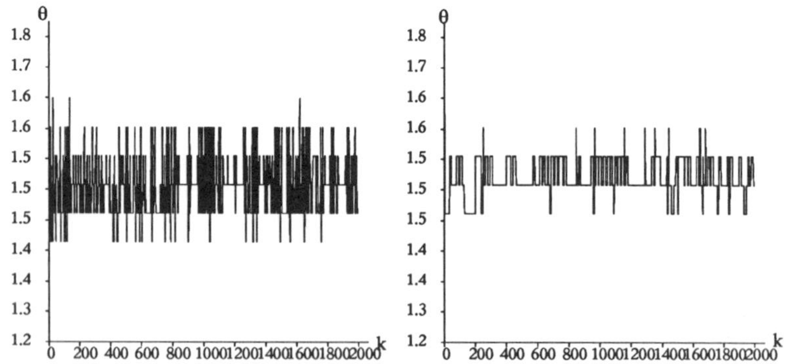

(a) $\sigma_\delta^2 = 0.00001$ (b) $\sigma_\delta^2 = 0.001$

Figure 7.9: *Plots showing the variation in bearing θ with noise model σ_δ^2 for a case with no relative motion between the feature and the sensor. With $\sigma_\delta^2 = 0.0000001$ (not shown) the feature was lost after about 500 steps. Variance of the process noise fixed at $\sigma_u^2 = 0.002$*

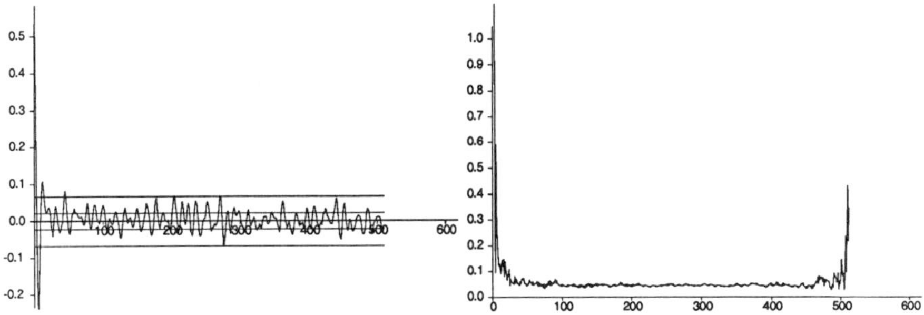

(a) (b)

Figure 7.10: *(a) Autocorrelation for the orientation during tracking of a stationary feature, showing* $\pm 3\sigma^2$ *and* $\pm\sigma^2$ *corresponding to whiteness confidences of 95% and 99%. (b) Power spectrum using a Bartlett window* $\Delta = 512$ *. Similar plots are obtained for the range of operational parameters giving satisfactory tracking performance*

the worst case that every return requires correctional-motion, the angular speed of the relative motion is limited to a maximum of approximately $0.6 rad/s$ tracking an RCD of width $0.17 rad$ for ranges of up to $4m$. However, this maximum speed can be increased to about $0.9 rad/s$ through the use of filter parameters, dead-band settings and a variable size of the correctional move step, optimized for a particular environment. Of secondary importance is the linear speed along the normal to the target. In addition to the firing speed the limiting factor in this case is the nature of the range validation and update scheme. Assuming no relative angular motion and with the range validation and update scheme as we have implemented it, the maximum linear speed along the normal to the target giving satisfactory results is approximately $2.5m/s$ [141].

Nature and range of the tracked feature. Within our operating range the tracking algorithm works better with diffractive targets i.e. edges and small cylinders. This can be explained from the fact that these features tend to have "better behaved" differentials i.e. actual differential characteristics which more closely approximate to the expected linear characteristic due to the diffractive nature of the received echo [139]. At short range $(0.45 - 1.5m)$, the data rate can be as high as $30Hz$ and at ranges of about $4m$ the maximum firing speed giving satisfactory results is about $25Hz$. And secondly, for a fixed separation d the ability to measure the differential accurately deteriorates with range.

Hardware Limitations. With the current hardware the differential can only be measured to a precision of $3.2\mu s$ up to a maximum of $15 \times 3.2\mu s$. The smallest step possible with the servo is $0.047rad$ which is crude given that the minimum RCD width tracked is $0.17rad$. However, these limitations can be overcome with improved hardware designs.

7.4 Modular Navigation

We now outline, as an application of the Tracking Sonar, a modular localization algorithm based on the non-linear *information filter*. The basis of most localization techniques is the use of beacons with respect to which location can be computed [54][31][5]. Representative of the state of art in navigation without artificially-setup beacons, is the approach by Leonard and Durrant-Whyte [133] on which the work presented here is based. It has been shown [133][139] that localization can be greatly improved by the use of sensor with the ability to maintain continuous map contact, i.e. sensors able to focus attention on known features in the environment as the robot moves. Having developed the Tracking Sonar with precisely such capabilities, we exploit the sensor for localization purposes as we now describe.

7.4.1 Localization Algorithm

The following are the appropriate states, observation and state transition models required to implement the information filter [140].

State. We define the location of the sensor platform in the environment as the state

$$\mathbf{x}(k) = [x(k), y(k), \alpha(k)]^T \tag{7.3}$$

where x and y are the Cartesian coordinates of the platform location and α is its orientation. An implicit assumption here is that the location of the sensing devices on the platform coincide with centre of the platform indicated by $\mathbf{x}(k)$ above. If however there is an offset a simple transformation can be used to take account of this.

Observations. The sensor makes an observation at time k of the parameters r and θ for a given RCD target whose model t is known. The observation vector $\mathbf{z}(k) = [r(k), \theta(k)]^T$, is a function of the state and the model t of the feature being observed. Define h such that:

$$\mathbf{z}(k) = \mathbf{h}[k, t, \mathbf{x}(k)] + \mathbf{v}(k). \tag{7.4}$$

It must be noted that unlike the algorithm of Leonard [132], we make use of *both* the range and the orientation in the localization cycle.

[5]There are alternatives to this such as the use of *occupancy grids* described by Moravec [145] and Elfes [61].

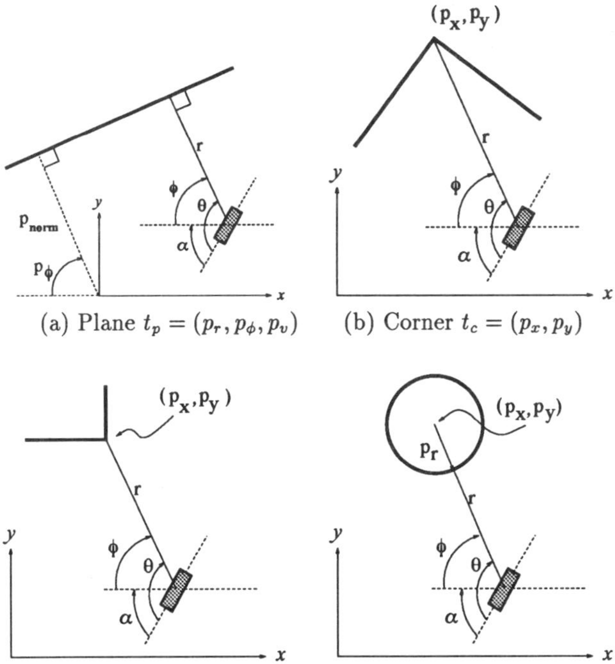

(a) Plane $t_p = (p_r, p_\phi, p_v)$ (b) Corner $t_c = (p_x, p_y)$

(c) Edge $t_c = t_e$ (d) Cylinder $t_{cyl} = (p_x, p_y, p_r)$

Figure 7.11: *Feature models in global coordinate system*

State Transition. The state transition equation is given by

$$\mathbf{x}(k+1) = \mathbf{f}[k, \mathbf{x}(k), \mathbf{u}(k)] + \mathbf{w}(k) \tag{7.5}$$

where $\mathbf{w}(k)$ is the noise associated with the transition and is defined by the covariance matrix $\mathbf{Q}(k)$. In the above equation the control input $\mathbf{u}(k)$ represents the motion of the sensor platform. Using a point kinematic model, we can write the control equation:

$$\mathbf{u}(k) = [T(k), \Delta\alpha(k)]^T \tag{7.6}$$

where the function $\mathbf{f}$ is determined by the system model, $T(k)$ is the linear component of the motion at time-step k and $\Delta\alpha(k)$ is the rotational motion at time-step k.

Implementation.

Assumption 1: A static two-dimensional world consisting of the four feature types described in figure 7.11.

Assumption 2: The existence of an accurate map and a corresponding RCD matching facility which matches the extracted RCDs and those predicted from the environment map [124].

The implementation proceeds as follows. The tracking sonar scans and extracts RCDs in its field of view and attempts to make a match. In a simple strategy the best matched RCD is selected. The parameters of the selected RCD can be used to initialize the information filter and the tracking sonar algorithm which are executed in parallel. An important consequence of this algorithm is that the estimates $\hat{\mathbf{x}}(k \mid k)$ are obtained at the sonar return rate.

7.4.2 Localization Results

The following results illustrate the performance of the localization algorithm. The estimates shown in all the results that follow were obtained in real-time at a data rate of approximately $30Hz$.

The results of Figure 7.12 show location estimates for motion in the x-axis and y-axis for two observation noise models. Other types of motion such as pure rotations and linear and rotational motion are also possible as long as the control input $\mathbf{u}(k)$ of Equation 7.6 is given. Figure 7.13 shows some results of localization estimates for motion which includes rotation.

In order to evaluate the positional accuracy of the localization, estimates can be compared with hand measured locations. In the results that follow, the platform moves from a known position B to a known position C and then back to B. The path followed between the points B and C is arbitrary in each case. The sensor platform stops at C for a known fixed time. In Figure 7.14(b) where the observation noise is modelled as being relatively large the absolute error between the estimated positions and hand-measured ones at C and B is quite large. According to the estimates of Figure 7.14 (b) and (d) the platform never reaches C and does not return to the original position B. In practice the actual relative level of the observation noise model chosen depends on such factors as speed of the motion and the positional accuracy with which the feature used for localization is matched.

7.5 Discussion and Summary

What we have developed and presented in this chapter is a model for a Tracking Sonar which is able to maintain focus of attention on a specified feature in real-time. There is ample room for further development of the hardware itself and optimization for various scenarios and applications. We leave the improvement of this technique to further research. However, even in this prototype phase the differential sonar can be made use of effectively in its various modes of operation. We have shown how the sensor can be used for vehicle localization providing estimates at rates of approximately $30Hz$. This is a vast improvement on the estimation speeds normally

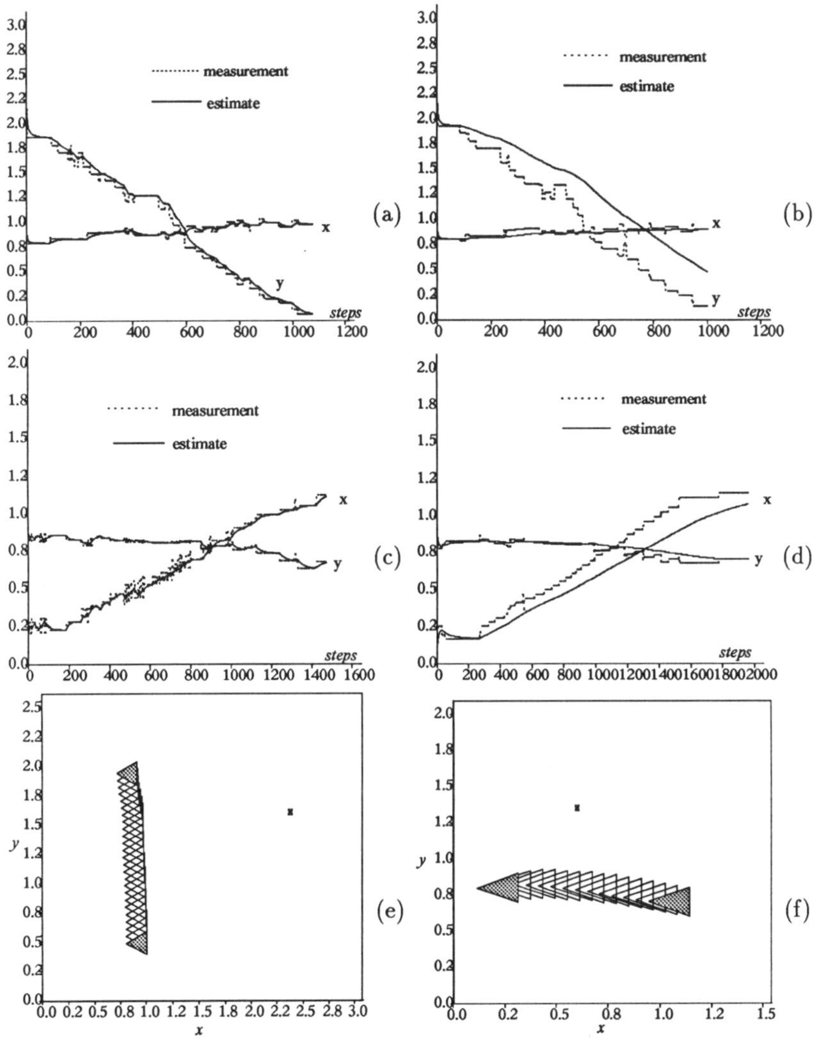

Figure 7.12: *Localization for motion in x-axis and y-axis for 2 observation noise models. In (a) and (b) motion is in the y-axis and in (c) and (d) motion is in the x-axis. In (a) and (c)* $\sigma_r^2 = 0.005$, $\sigma_\theta^2 = 0.01$, *(b) and (d)* $\sigma_r^2 = 0.05$, $\sigma_\theta^2 = 0.5$. *The transition noise matrix* $\mathbf{Q}(k)$ *was fixed. (e) and (f) show the motion plotted in (b) and (d)*

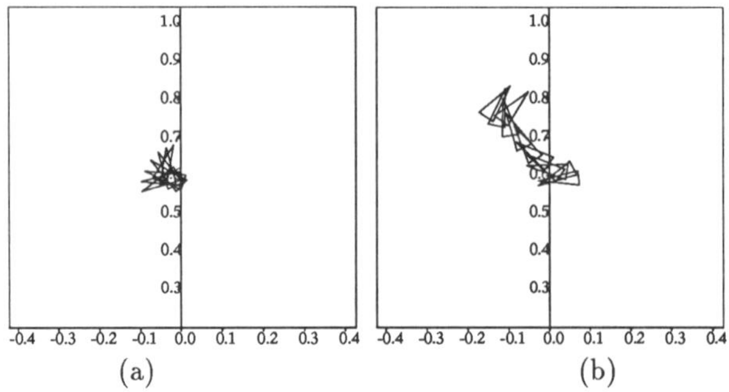

Figure 7.13: *Localization estimates for sensor platform incorporating rotational motion (a) pure rotation (b) linear and rotational motion*

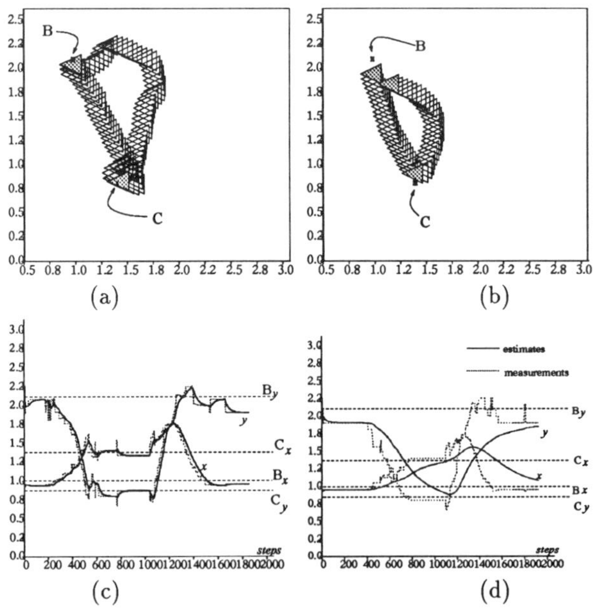

Figure 7.14: *Comparing estimates with hand-measured locations. Motion from B to C and back to B stopping at C and B for a fixed period (10 seconds). The true (hand-measured) positions B and C are indicated. (a) and (b) show the estimated location of the platform. (c) and (d) show x y positions calculated directly from observed r and θ and the estimated x and y from the localization information filter. In (a) and (c) the variances in* $\mathbf{R}[k, \mathbf{z}(k)]$ *are* $\sigma_r^2 = 0.0001$, $\sigma_\theta^2 = 0.001$, *in (b) and (d)* $\sigma_r^2 = 0.05$, $\sigma_\theta^2 = 0.5$

achieved using sonar [133] as a result of overcoming the ' stop-look-move' handicap. The ability to localize can be used as a basis for classifying features as discussed in [139].

Use can be made of the Tracking Sonar in a environment where other similar sensors are operating. When such sensors are used to track different features which are not in close proximity, the performance is the same as in the discussion above for a single sensor. However, when the sensors are used to track the same feature or features in very close proximity ($\approx 0.2m$ apart), interference may occur if very high data rates of the order of $30Hz$ are maintained. However, reliable operation can be obtained with data rates of approximately $20Hz$. The ability to use several Tracking sonars together means that a decentralized localization scheme can be developed. In such a scheme sensors are configured as illustrated in Figure 7.15. A decentralized

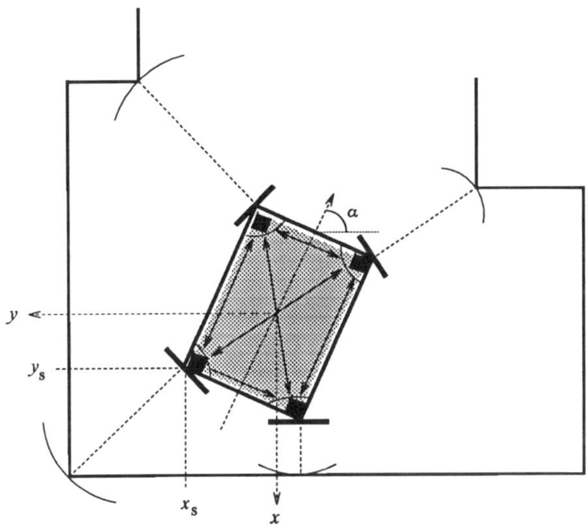

Figure 7.15: *Tracking Sonar configuration for another of the Oxford vehicles (namely the modular OxNav vehicle). Each sensor focusses attention on a particular feature of the environment for purposes of localization or map-building. The sensors are decentralized and fully connected*

form of the non-linear information filter (see the Appendix which follows) [139] can be used to generate the global estimate of the vehicle's location. Work is currently in progress on such a decentralized localization scheme.

Chapter 8

Architectures and Algorithms for 3-D Vision: The Parallel Droid System

Han Wang, Michael Brady and Chris Bowman

This chapter describes the implementation of a 3D vision algorithm, Droid, on the Oxford parallel vision architecture, PARADOX, and the results of experiments to gauge the algorithm's effectiveness in providing navigation data for an autonomous guided vehicle. The algorithm reconstructs 3D structure by analysing image sequences obtained from a single moving camera. The primary components of PARADOX are a Datacube pipelined image processor, a configurable network of 32 T800 transputers, a Sun-4 workstation and a special-purpose interface connecting the Datacube to the transputer network. The Datacube handles the image digitisation, storage and display; the transputer network performs the feature extraction (corner points) in parallel and the Sun-4 computes the 3D construction. In this application, the architecture delivers a performance of greater than 1 frame per second – 17 times the performance of a Sun-4 alone. Experiments suggest that this system is capable of maintaining an estimate of true camera position to within 5% over 600 image frames by using visual data alone.

This work is a step towards the eventual goal of the vision group in the AGV project, which is to use passive vision in the real time control loop of the vehicle. Currently the Droid project demonstrates the potential in navigation.

8.1 Introduction

PARADOX [192] is a hybrid parallel architecture which has been commissioned at Oxford in order to improve the execution speed of vision algorithms and to facilitate their investigation in time-critical applications such as autonomous vehicle guidance. Droid[88, 87, 49, 186, 20] is a ***structure-from-motion*** vision algorithm which estimates 3-dimensional scene structure from an analysis of passive image sequences taken from a moving camera. The motion of the camera (*ego-motion*) is unconstrained, and so is the structure of the viewed scene.

Until recently, because of the large amount of computation required, Droid has been applied off-line using prerecorded image sequences, thus making real-time evaluation of performance difficult. In this experiment, the video camera has been mounted on the AGV, and the Droid algorithm implemented on PARADOX.

This chapter outlines some of the key features of the Droid algorithm (see [88] for more detail), the PARADOX architecture and the parallel implementation of Droid on PARADOX using a mixed programming approach. The performance of the system is reported in terms of execution speed and accuracy of ego-motion and vehicle-speed estimation. The latter are compared with the values obtained using the on-vehicle laser scanning navigation system.

8.2 Vision Algorithm: Droid

Droid is an algorithm developed for recovering structure from motion. Assuming that the camera is in motion in an otherwise static environment, and that the input is a sequence of digitised image frames, structure from motion algorithms use the observation that the movement of a feature in the image between different image frames depends on both the movement of the camera and the feature position. For example a vertical edge to the left of the camera moves further to the left in the image as the camera moves towards it. In particular structure from motion algorithms establish the following: (i) the correspondence between features in temporal space and (ii) the camera motion (also referred to as ego, self motion). The ego motion has six degrees of freedom: three in translation and three in rotation. Since there is a speed-scale ambiguity, the ego motion can be recovered from the correpondence of only five points. In the Droid algorithm described in this paper, the ambiguity is overcome by introducing a scale at the booting phase for the first two frames.

Droid functions by detecting and tracking discrete image features through the image sequence, and determining from their image-plane trajectories both their 3D locations and the 3D motion of the camera. The extracted image features are assumed to be the projection of objective 3D features. Successive observations of an image feature are combined by use of a Kalman filter to provide optimum 3D positional accuracy.

The image features originally used by Droid are determined from the image, I, by forming at each pixel location the 2×2 matrix, $A = w \otimes [(\nabla I)(\nabla I)^T]$, where w is a Gaussian smoothing mask. Feature points are placed at maxima of the response function R [88]

$$R = det(A) - k(trace(A))^2 \tag{8.1}$$

where k is a weighting constant. We understand the significance of the terms in this expression as follows. The eigenvalues of A are proportional to the principal curvature and are rotationally invariant. That is to say, if both curvatures are high and reach local maxima (which we can interpret as corners) the product of both eigenvalues is high. It is not difficult to show that the $det(A)$ is the product of both eigenvalues; therefore large values of $det(A)$ may be used to predict corners.

The second term $trace(A)$ in equation 8.1 is the edge reponse and is used to suppress false corner responses on the edge. Often, features are located near image corners, and so the operator tends to be referred to as a corner finder. In fact, it also responds to local textural variations in the grey-level surface where there are no extracted edges. Such features arise naturally in unstructured environments such as natural scenes.

Manipulation and matching of corners are quite straightforward by using the attributes of grey scale, x-gradient and y-gradient. More reliable matching algorithms exist, but at a cost, such as the grey level convolution. The use of a Kalman filter reduced the burden of matching by 'intelligently' predicting the next matching candidate. In the current implementation, the depth map is constructed from tracked 3D points using a local interpolation scheme based on Delaunay triangulation [68].

Droid runs in two stages: the first stage is the booting stage, called boot mode, in which Droid uses the first two images to start the matching process; the second stage is the run stage called run mode.

1. Boot mode

In the boot mode, points in the two 2D images are matched using epipolar constraints. The matched points provide disparity information which is then used for estimation of ego-motion and 3D instantiation. Ego-motion is described as a 6-vector (3 in translation and 3 in rotation). Each matched point gives one constraint on the ego motion. In principle then, 6 points are required to resolve the camera motion. However, there is a speed-scale ambiguity; the scale of the translation is indeterminate from the visual data alone. To remove this ambiguity a prior estimate of camera motion must be given at booting phase. Hence at least 5 point matches and a prior estimate are sufficient to determine the ego-motion. In practice, the extra constraints introduced by further point matches are used to improve the conditioning of the ego-motion calculation.

2. Droid run mode

The run mode of Droid includes the following: (i) a 3D-2D match which associates current 3D points with the newly detected 2D points, (ii) an updated ego-motion estimation and (iii) a 2D-2D match, between residual points in the feature points list and unmatched points from the previous frame, to identify new 3D features. Also, 3D points which have been unmatched over a period are retried in this phase.

In summary, the Droid algorithm consists of the following stages:

- corner detection
- refining of existing 3D corner position estimates by matching to corresponding new 2D feature points
- estimation of ego-motion
- instantiation of new 3D points in space by matching unmatched 2D points to unmatched 2D *limbo* points from the previous frame
- retiring of unmatched 3D points
- 3D scene reconstruction (interpolation)

For the purposes of this parallel implementation, the Droid algorithm has been broken into two parts: corner detection and the rest. In fact, corner detection is a data independent operation which accounts for more than 90% of the entire computation on a serial processor. The remainder of the algorithm is data dependent with relatively higher-level data structures. In our parallel implementation, the second part will be referred to as 3D construction.

8.3 PARADOX Architecture

As we discuss in chapter 5, a major problem in using passive vision is bandwidth, both in acquiring and processing data. In this section we describe PARADOX, a hybrid architecture, designed and configured especially for vision/image processing algorithms. It consists of three major functional parts: a Datacube pipelined system, a transputer network and a Sun4 workstation. The architecture of PARADOX, as applied to Droid, is shown in Figure 8.1. This section describes the functionality, connectivity and programming of its components.

The Datacube family contains more than 20 types of VME-based pipelined processing and input/output modules which can perform a wide range of image processing operations at video rates. Image data is passed between modules via a Datacube bus-standard known as the MaxBus. System control is by means of the VME bus from the host Sun workstation. The Datacube can be used for image digitisation, storage and display and also for a wide range of video frame rate pixel-based processing operations.

The transputer network consists of a fully populated Transtech MCP 1000 board. This contains 32 T800 transputers – each with one Mbyte of RAM – and both hard-wired and programmable switching devices to allow network topology to be altered.

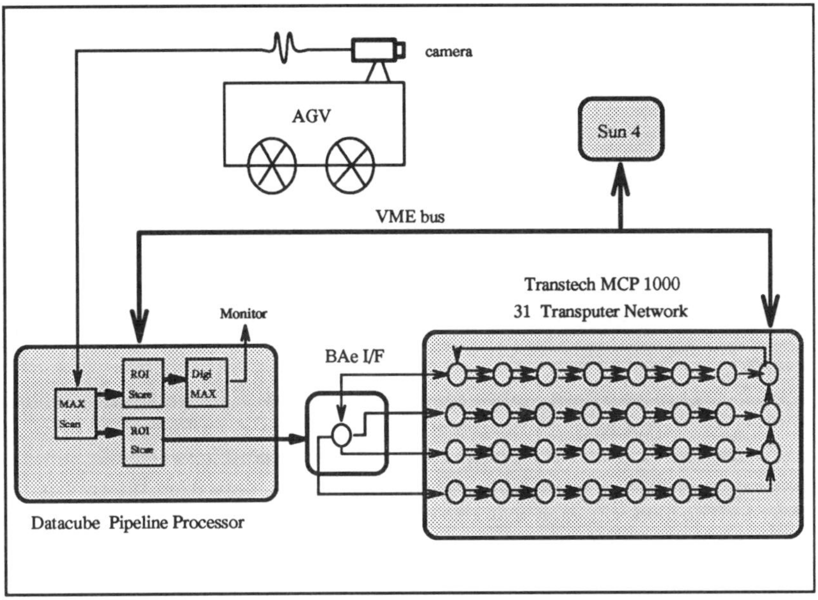

Figure 8.1: *Machine architecture of PARADOX (Droid incarnation)*

A wide range of network topologies can be implemented including parallel one dimensional arrays [193], a 2D array or a ring structure. This board delivers a peak performance of 320 MIPS.

The connection between the Datacube and the transputer network is by way of the BAe interface, an interface board designed by the British Aerospace Sowerby Research Centre. This card has a 512 Kbyte dual-ported video memory which can be accessed via the Maxbus and also by an on-board T800 transputer. The video memory is double buffered so that 512×512 images can be transferred to and from the Datacube in a video frame period while the T800 is accessing other parts of the memory. The BAe interface passes image data to and from the transputer network in parallel via the four transputer serial links. The four links can sustain a data transfer rate sufficient to allow video-rate transmission of 256×256 images and it is this that determines the throughput limit of the interface. Control of the BAe interface is implemented by loading code into the on-board T800 from the Sun using the transputer link structure.

Figure 8.2 depicts the hierarchy in programming PARADOX. As might be expected with such a hybrid structure, the programming of PARADOX is a mixture of various programming methods, ranging from low-level register setting to complex algorithms, and languages. Generally, we classify the programming of PARADOX into three levels of hierarchy:

(i) Level 1
This level consists of the lowest-level functional modules for programming PARADOX. *Configuration* refers to the physical process of interconnecting the data and control pathways in the architecture in both the Datacube and the transputer array. For example, in the transputer array, links between processors are connected under programme control using the on-board cross bar switch. The *Networking Library* is a set of routines for data communication between the transputer network and the Sun workstation using a predefined *Network Protocol.* The *Occam Harness* provides a user interface for level 2.

(ii) Level 2
Level 2 encompasses both *system control* and a library of *image processing* functions. The control routines in *system control* coordinate computations between the Datacube, Sun and the transputer array. For example, when the Sun has finished one frame, a control process will read in new data from the transputer array for the next iteration and also command the transputer network to begin computation on the next frame.

(iii) Level 3
Level 3 is the application level. At this level the *vision application* algorithms utilise the low level image processing routines and the system control functions. Data structures at this level are the most sophisticated. Droid (see next section) is implemented at this level.

At Level 1, most programs are written in Occam; at Level 2, programs are written in C, parallel C and Occam; while at Level 3, algorithms are usually written in sequential and/or parallel C.

8.4 Implementation of Droid on PARADOX

In the parallel implementation of Droid [192], the Datacube is used to digitise, store and display the image sequences and graphics overlays; the corner detection is carried out by the transputer array and the 3D construction is computed on the Sun workstation.

8.4.1 Datacube

In the Datacube, a MaxScan board is configured and programmed to digitise 256×256 grey level images from the camera mounted on the robot vehicle. These images are simultaneously stored in buffers in two region of interest-stores. One region of interest-store acts as a circular image buffer and displays, via the Digimax, a previously stored

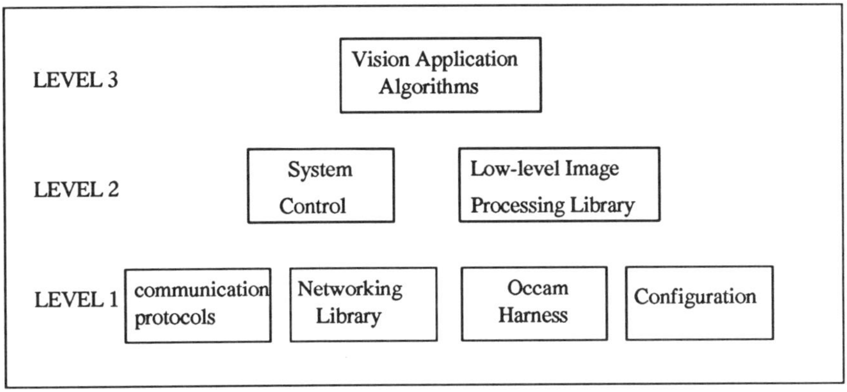

Figure 8.2: *Hierarchy in programming PARADOX*

frame together with the Sun generated overlays representing the inferred 3D geometric data output from Droid for that frame. The other region of interest-store repetitively broadcasts each frame in a form suitable for reception by the BAe interface.

Under live conditions, the synchronisation of PARADOX and the timing of Droid are important issues. Droid requires each frame to be tagged with a timing "stamp". This presents some problems as the control path between the Sun (which directly controls the Datacube) and BAe interface is via the transputer network and the timing delays through the latter are not easily predetermined. This problem is overcome by using the Datacube in a two stage mode: a single video frame is first digitised into a region of interest-store and then this region of interest-store broadcasts the frame continuously onto its data bus. In the mean time, a tag has been sent to the BAe interface board, via the transputer network, to inform that this frame is ready to be captured. The frame will then be stored into BAe's video RAM where it can be accessed by the on-board T800, together with the time stamp read from BAe's local clock.

8.4.2 The Transputer Network

A corner is detected by computing the local response function R in equation (8.1). If r denote the computing process of corner detection, the input of r is the grey-level image and the output is a list of corners. Decomposing the image into n partitions, there exists n parallel processes by replicating r. Figure 8.3 depicts the process r and its replication. On an MIMD computer, such as a transputer network, the replication can be most easily accomplished by mapping each process r onto one processor. The throughput of such a system is n times greater than a single processor minus the input/output overhead associated with partitioning and delivering images [193, 72].

Functionally, the transputer network, including the BAe interface T800, is con-

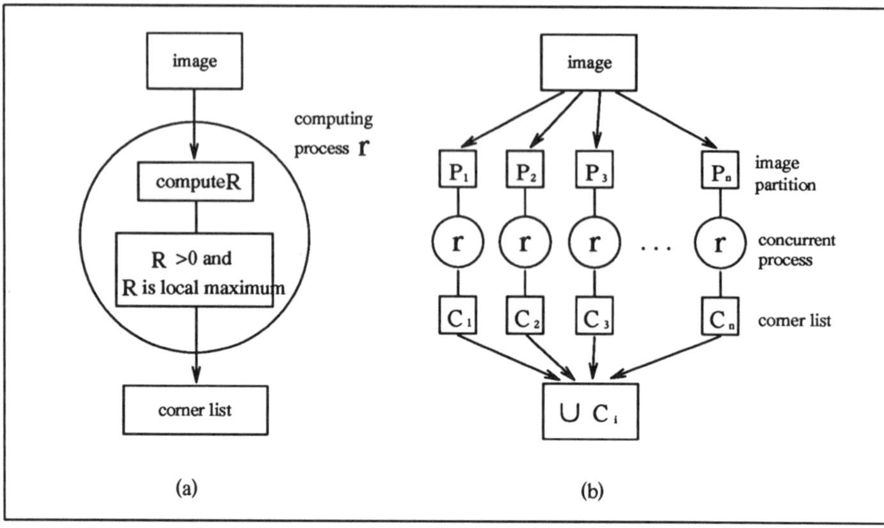

Figure 8.3: *Partitioning of the input image and replication of the corner detection process r*

figured as a multiple one-dimensional array with three functional parts: (1) a server node, (2) an image grabbing node, and (3) computing nodes. The server node acts as a communication interface between the Sun and the transputer network. It passes commands from the host to the image grabbing node on the BAe interface. The extracted corner features are also passed back through the server node into the Sun 4. The image grabbing node partitions the image into 4×7 segments and transmits them in parallel to the network, where each computing node intercepts a segment and computes the corner locations. The other transputers on the board are dedicated to data routing.

8.4.3 Control Flow in Droid on PARADOX

Figure 8.4 shows the flow of control of Droid on PARADOX. The signal $s1$ initiates the execution. $p0$ represents the boot mode, $p1$ represents the run mode of 3D construction, $p2$ is the grabbing process of the Datacube, r is the corner detection process, and $t0, t1$ are timing signals.

Once the starting signal $s1$ is issued, video signals from the camera are digitised and broadcast to the Maxbus on the command $g0$ from the Sun 4 at every Droid iteration. At each iteration, process $p1$ receives data $d2$ which includes the corner point list and timing information. At the same time odometry data from the GEC laser-scanner navigation data is obtained via a serial link between the on-vehicle

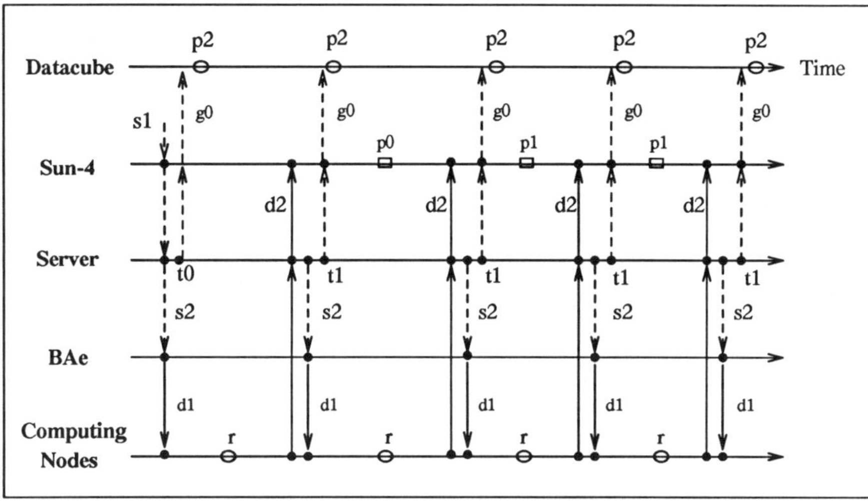

Figure 8.4: *Control flow in Droid on PARADOX*

control board and the Sun 4 (note that this is used to assess the performace of the Droid algorithm, and is not required in the algorithm itself). The computing nodes receive grey-level image partitions $d1$, sent by the BAe board, and return detected corner points.

8.5 Performance Evaluation

Figure 8.5, top, shows an image of a corridor in the laboratory. Corners detected by the Droid corner finder are superimposed. Figure 8.5, bottom, shows the same image with a Cartesian grid plot of the interpreted 3D surface by Droid superimposed. The driveable region can be clearly identified. In another run obstacles (three rubbish bins) were placed in the driving path and were identified as shown in figure 8.6. An algorithm has been developed by D. Charnley [49] to extract the drivable region by computing the surface normal of each grid.

The above demonstrates qualitatively the performance of Droid in live situations, but not quantitatively. A series of experiments has been conducted at Oxford and at Plessey (Roke Manor) to measure the performance of Droid in both live and static environments. The intention has been to demonstrate the competence of dynamic vision in a real environment. The following sections describe some of the results.

8.5.1 Execution Speed

The performance obtained from PARADOX for parallel Droid was 0.87 seconds per frame. This is 17 times faster than a pure Sun-4 implementation. The overall performance is limited primarily by the parallel execution of the two phases of Droid: the 3D construction and corner detection algorithms which have comparable execution times in the parallel architecture (compare this with the serial architecture in which the corner detection takse 90% of the time). The Datacube control and visual display functions contribute a negligible fraction of the execution time.

This performance could be improved by off-loading some of the initial corner detection operations, such as global convolutions, onto the Datacube modules which perform a hardware implementation. However, the Droid algorithm requires 16-bit fixed-point precision fairly early in the corner detection algorithm and this requirement limits the possibilities. Another approach would be to parallelise the 3D construction software and transfer some parts to the transputer network. Both of these are being considered at present.

Recently, an improved corner detection algorithm has been developed and is under test at Oxford [194]. This uses second order directional derivatives with the direction tangential to an edge. This algorithm has improved accuracy of corner localisation and reduced computational complexity. Consequently, it allows faster execution speed (7 frames per second) than the original Droid corner detection algorithm. This, together with parallelisation of the 3D construction algorithms, will offer further improvements to overall execution speed.

8.5.2 Ego-motion and Vehicle Speed Estimation

The laser scanner on the vehicle determines the AGV's location (2D position and orientation) by detecting fixed bar-coded navigation beacons (Chapter 2). Acquiring this data at each step allows us to compare its predictions (which are known to be accurate and which we shall call the the "true" AGV trajectory) with that predicted by Droid. The following results were obtained from an experiment where the AGV was programmed to run in a straight line with varying speeds.

Figure 8.7 depicts a plane view of the AGV's trajectories: the solid line represents the AGV trajectory reported by Droid and the dashed line as reported by the laser scanner. In this particular run, the AGV has been programmed to move in a straight line at two different speeds. For the first part of the run it travels at about 8 cm/sec and for the second it travels at about 4 cm/sec. Droid reports the camera position (6 degrees of freedom – 3 translation and 3 rotation) from the starting point of the vehicle, which has coordinates $(x_0, z_0) = (0, 0)$ and the laser reported trajectory is re-aligned accordingly. Figure 8.7 shows that the alignment between the laser scanner readouts and the Droid prediction suffers from drift over distance, but diverges by only a few per cent over 10m.

Figure 8.8 shows more clearly the error between the Droid predicted position and the laser readings. After a few hundred frames, a progressive drift of the Droid

Sequence	Number of frames	Error in rotation(y) unit: rad	Error in translation(z) unit: m	Total translation unit: m
1	141	0.0363	0.63	7.31
2	155	0.0453	0.32	8.11
3	140	0.0555	0.27	8.19
4	91	0.0238	0.11	4.42
5	143	0.0311	0.09	7.35
6	177	0.0899	0.61	8.92
7	180	0.1066	0.17	8.95

Table 8.1: Measurements of Droid Performance

reported position is evident. Obviously, this is because Droid refers to its position with respect to its starting place and errors accumulate. This problem could be resolved by using the laser-scanning navigation system as a reference (Chapter 3) and periodically aligning Droid's position with this.

During the run, the vehicle has been stopped twice manually to test this system's tolerance under different situations. Figure 8.9 shows the speed of the AGV as determined by Droid (solid line) and by the on-board GEM80 using the laser scanner (dashed line). The speed plots in figure 8.9 agree closely apart from the moment when the vehicle alters its speed where Droid consistently overshoots. This can be improved using non-critical damping.

Table 8.1 summarizes performance figures of 7 executions drawn from more than 40 separate runs. Errors computed by Droid are measured with reference to the laser scanner readings.

Droid yields ego motion that have 6 degrees of freedom. However the laser scanner can only give planar measurements. Hence, comparable measurements are only the rotation about y axis and translation in total distance (Euclidean distance on the X-Z plane). On average, translation error of Droid is 4.0% (standard deviation 1.0%). The mean error of rotation about y axis is 0.05 radian. Errors of rotation on other two axes are larger than on the y axis. This can be explained as the following.

Because of speed-scale ambiguity, Droid's estimates of ego-motion are critically dependent on a pre-estimate of motion during the booting phase. When operating from "canned" image sequences, the translation between the first two images can be accurately determined a priori and motion scale can also be re-adjusted at any point. However, in live mode and without other navigational input, this becomes impossible. Under (or over) estimation at booting phase will result in under (or over) prediction during the running phase. Improvements are under investigation. For example, a twin camera system and stereo analysis would provide more accurate ego motion estimation under most conditions. Another improvement would be to improve the timing accuracy of the image grabbing by the BAe board.

8.6 Conclusion and Future Work

Droid constructs an explicit three-dimensional representation from feature points extracted from a sequence of images taken by a moving camera. This chapter has described the algorithm, the PARADOX parallel vision architecture and the implementation of Droid on PARADOX. Experiments have demonstrated the competence of Droid and the performance of PARADOX in dealing with real world problems. The results show that this system is capable of identifying basic surface structure and can be used to supply purely passive guidance information for autonomous vehicles where other sensory mechanisms find it hard or impossible.

Future work includes (1) the incorporation of the fast corner detection algorithm developed at Oxford into Droid [194, 30], (2) the use of odometry information taken from the AGV to provide Droid with more accurate motion estimations, and (3) the final goal: to close the control loop of the AGV using passive vision, the information provided by Droid.

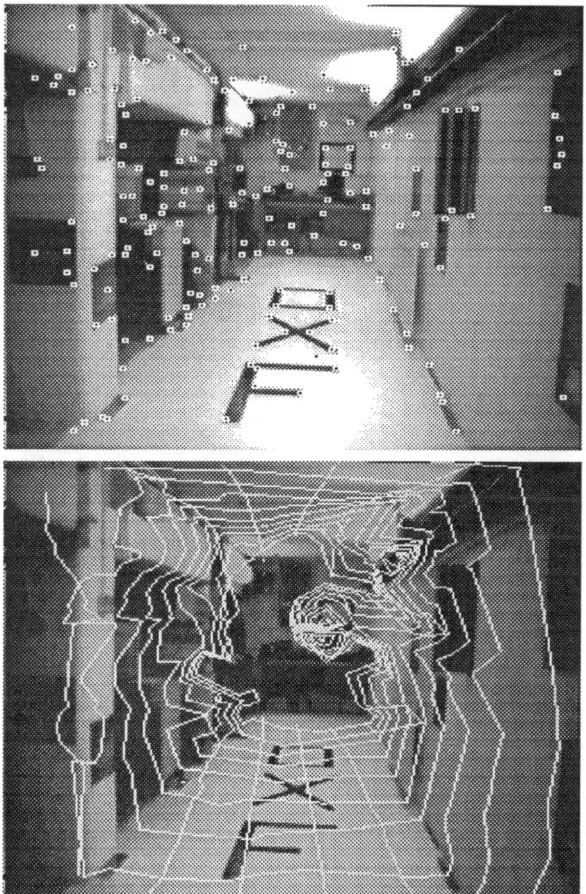

Figure 8.5: *Top: One grey level image with corners superimposed, taken from a sequence of 200 frames; Bottom: A Droid reconstructed 3D surface. A driveable region can be clearly identified in the bottom picture*

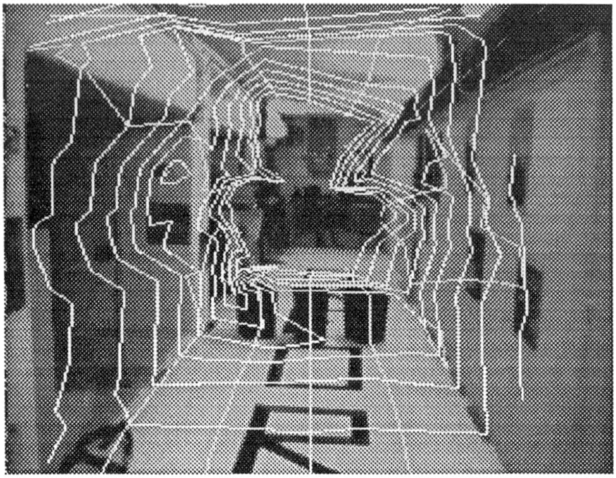

Figure 8.6: *Reconstructed surface with obstacles on the driving path*

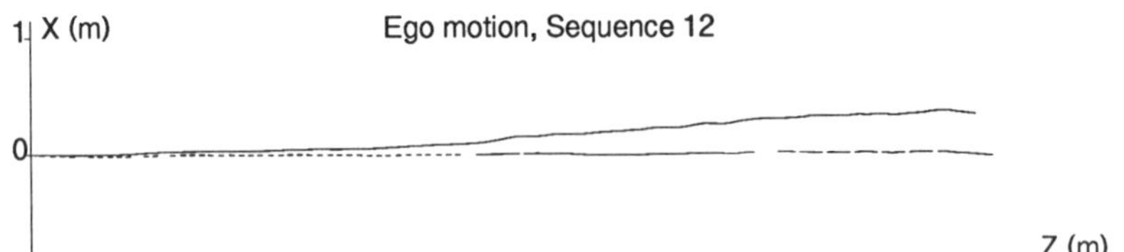

Figure 8.7: *Plane view of AGV trajectory. Solid line–Droid predicted motion; Dashed line–laser scanner readings*

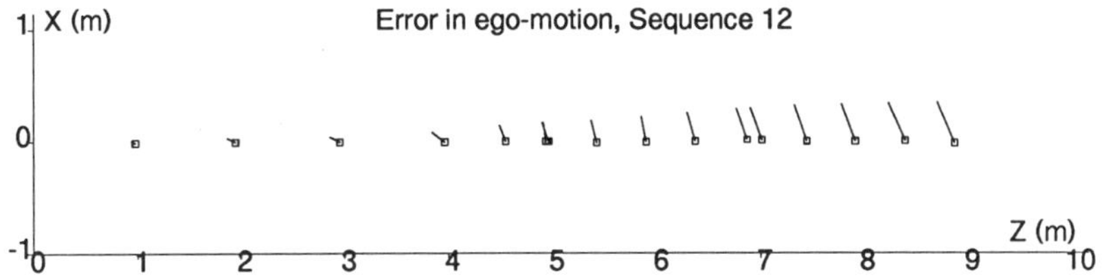

Figure 8.8: *Errors of Droid predicted AGV trajectory using the laser scanner reported positions as reference*

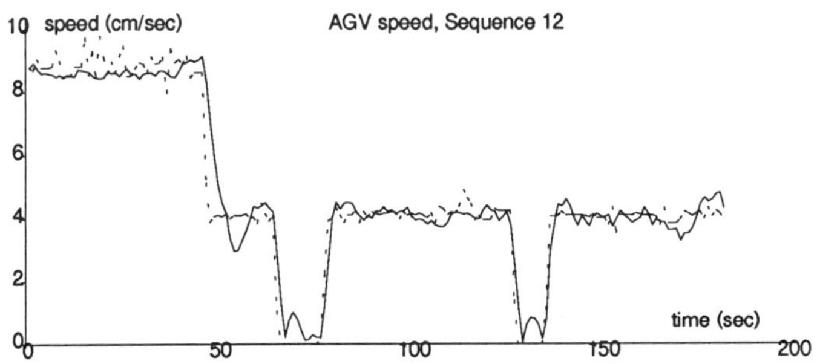

Figure 8.9: *Comparison of AGV speed. Solid line–speed reported using Droid; Dashed line–speed reported using laser scanner*

Chapter 9

Range Image Feature Extraction and Representation

Ian D. Reid and J. Michael Brady

Consider a mobile robot performing a task in an unknown or partially known environment. If it is to react to that environment in other than a haphazard "trial-and-error" fashion, the robot must learn about its surroundings using sensed data. Perhaps the simplest requirement is to avoid obstacles; for this task the robot need know nothing more about obstacles than roughly their positions and extents. Sensor directed obstacle avoidance is discussed elsewhere in this book. A considerably more complicated task involves finding and manipulating a *specific* object in the environment. In this case the robot must not only locate the object, but also *recognize* it. Recognition demands the use of both sensed data and *a priori* knowledge about the object in order to detect its presence, and to determine its pose (position and orientation) relative to the sensor.

The objects of interest in material transfer applications of relevance to the Oxford AGV project tend to be at distances from 1 – 10 metres from the vehicle, and they tend to be instances of a limited set of object classes. For example, pallets are all similar in shape and construction, though the number of slats, the height of the pallet, and the aspect ratio, vary considerably from pallet to pallet. Short-cuts, such as attaching retro-reflecting markers like bar codes to pallets, etc., were ruled out by our industrial collaborators as being too restrictive for their needs. We explored the use of a variety of sensors for object recognition and pose determination, and concluded that range imaging was the best strategy for the short term. Intensity images from passive CCD cameras lose depth and orientation information through the projective transformation of image formation. For object recognition, range information needs to be dense and three- dimensional. Sonar is too slow and its spatial resolution too poor, for recognizing objects such as pallets and containers; therefore we use optical techniques.

Range data as a cue for recognition has the virtue that it preserves 3D properties of features in the environment such as surface orientations and positions. One particularly useful form of range data is the *range image*, a dense array of values related to scene distance. Such an image contains much geometric information, in the form of *image features* such as edges and surfaces, which can be extracted from the image and used via model-based vision techniques to give direct constraints on the recognition process. It is interesting to compare the description of these 3D features with the sonar based 2D equivalent features described in Chapter 7.

In this chapter we discuss the utility of various range image features as cues for model-based recognition to drive the task determination level in the AGV controller. In particular we are concerned with robust, efficient extraction of features which will not only aid recognition and localization, but will provide maximal constraining power on unknown parameters of a model class. We derive parameters from first and second order feature matches using basic tree searching matching algorithms. The emphasis is on using simple local features and strategic algortihms to determine model parameters, resulting in a efficient method for object recognition.

Range data is acquired from a novel laser stripe triangulation system, the Oxford/NEL laser range finder (G. T. Reid *et al.* [170] and I. D. Reid *et al.* [174]) which is fitted to the AGV vehicle and which supplies high density range images at moderate bandwidths over distances from 1.5 to 5 metres. This sensor differs from the simpler LEP triangulation sensor described in Chapter 6 in acquiring three dimensional data through scanning a stripe and triangulating onto a CCD array. We begin the chapter with a discussion of range image formation on this sensor, and consider the features available in such an image and methods for their extraction. There follows in Section 9.2 an analysis of the utility of each of these feature types with respect to recognition and, in particular, determining the internal parameters of an instance of an object class. This analysis leads to the formulation of a matching strategy for recognition and parameter determination, in Section 9.3.

9.1 NEL Range Images and Features

9.1.1 Image Formation

Figure 9.1 shows a photgraph of the Oxford/NEL range finder. The sensor is based around the well established idea of triangulation using structured light and a passive camera. A light stripe is formed using an infra-red diode laser spread by a cylindrical lens. The stripe is reflected into the scene by a mirror driven by a stepper motor. The novel feature of the design of this sensor is that the camera views the scene *in the same mirror*. This configuration achieves the same effect as if the entire sensor

Figure 9.1: A photograph looking down on the NEL range finder. The laser is located parallel to the bottom of the picture. The camera is located above this, and both are directed to face the mirror, to be found on the right of the picture

head were rotated, but is achieved rather more quickly and cheaply.

The sensor is equipped with special purpose "hit-box" hardware so it is not necessary to store images of the stripe and process them later. For each mirror position the contents of the camera's CCD array are processed by the hit-box which counts across each scan-line until it finds the stripe or reaches the end of the array. A stripe is detected by thresholding each scan-line of the detector array to isolate the stripe from other image features. The leading and trailing edges of the stripe are identified and stripe peak taken to be the average of the two. Vertical resolution is determined by the camera hardware; we use a standard interlaced CCD camera and ignore even fields giving a vertical resolution of 287 pixels. Each odd field gives rise to one column of data so the sensor has a bandwidth of approximately 15kHz. Columns are transmitted to Datacube hardware which acquires the data and aggregates columns into a range image. Horizontal resolution is determined under software control by varying the scanning speed of the mirror. A sample range image acquired using the sensor is shown in Figure 9.2.

9.1.2 Feature Interpretation and Extraction

Many researchers have considered the problems of feature extraction from range images (for example, Besl and Jain [16], Brady *et al.* [29], Faugeras and Hebert [65], Blake and Zisserman [19], Yokoya and Levine [197]) and the techniques we describe in this chapter have been well documented in the past. We present this work in the context of an evaluation of the tradeoffs involved between, on one hand, the features which are present in a range image and which can be reliably extracted, and on the other, those which are best and/or the most convenient for recognition. In the spirit

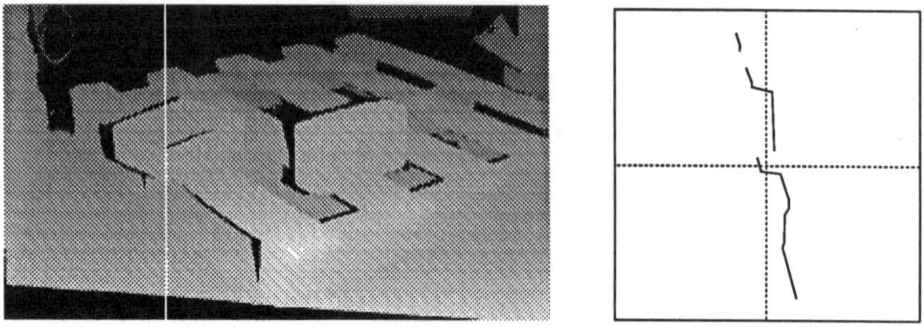

(a) (b)

Figure 9.2: Range image formation: (a) a sample NEL range image; (b) the appearance of the stripe on the CCD which gave rise to the indicated column of the image

of Bolles and Horaud [23], who analysed the features to be found on a single stripe from a triangulation range finder, we begin this chapter with a discussion of the features which appear in NEL range images of polyhedral scenes, and discuss the issues of how, and how reliably, these features can be extracted.

In the previous section we discussed the format of an NEL range image. The pixel values in such an image are related to depth, but not by a simple proportionality; i.e. the image is not a rectangular voxel representation. However, the main features which can be extracted from a range image have simple real-world interpretations:

- a planar patch in a range image corresponds to an area of planar surface;
- a step edge in a range image (with valid data on either side of the edge) corresponds to an occluding boundary, and hence the near side of the edge corresponds to a surface boundary;
- a crease edge in a range image corresponds to a join of surfaces with different orientations.

Patches, regions of the image surface, are the primary visible features, while steps and creases are first and second order discontinuities in the image surface, respectively.

Surface points

Grimson's system for recognition of 3D objects from 3D data uses a position/surface normal representation. Such data are readily extracted from a range image by finding a small patch of image points and estimating the surface normal from them. If the image surface is given by the function $I(x, y)$, then the normal at position (x, y) is the vector

$$(n_x^2 + n_y^2 + 1)^{-1/2} \begin{bmatrix} n_x \\ n_y \\ 1 \end{bmatrix}$$

where $n_x = \frac{\partial I}{\partial x}|_x$, and $n_y = \frac{\partial I}{\partial y}|_y$. Discrete masks for estimating these partial derivatives have been published in numerous places (see, for example Haralick [84]) and are not given here.

Patches and regions

The advantage of a patch-based over a point-based representation of the range data is that a patch contains information about the extent of the surface whereas a point only represents the plane equation of the surface.

Patches can be constructed by merging adjacent points whose positions and surface normals satisfy approximately the same plane equation. Haralick [84] has shown that under appropriate distribution and independence assumptions for the noise, the decision to merge is given by a test he calls the F-test.

In our work we first identify small planar patches. A simple method for extracting a number of these is to find regions of identically zero Laplacian. Let

$$I_{xx} = \frac{\partial^2 I}{\partial x^2}; \quad I_{yy} = \frac{\partial^2 I}{\partial y^2} \tag{9.1}$$

If one of I_{xx} and I_{yy} is zero as well as the Laplacian and the region has non-zero area it must be planar. The Laplacian can be approximated locally using the 3×3 mask below.

-1	-2	-1
-2	12	-2
-1	-2	-1

We have implemented a version of this algorithm on our Datacube pipelined processing system using a VFIR-II convolution board to estimate the Laplacian, followed by dilation of the Laplacian convolved image (a morphological operation, see Noble [150], implemented using SNAP board: Systolic Neighbourhood Array Processor). Points in the image falling below a threshold can then be marked as the centres of planar patches (implemented using look-up tables on the SNAP board).

Extended patches are then constructed by growing regions from these seed points until either a step-edge is encountered (see the following section) or the estimated surface normal at the current was sufficiently different from the estimated surface normal at the original point. An example is shown in Figure 9.3.

Steps

Step edges in a range image correspond to depth discontinuities or to the boundary between valid range data and no valid range data. In the former case they always mark surface boundaries, and in the latter case they often mark surface boundaries.

Figure 9.3: An NEL range image of a pallet and the planar patches extracted: the patches are indicated in black, with white meaning no valid range data

It is beyond the scope of this discussion to review the plethora of techniques proposed for finding step-edges in intensity images. The key observation here is that the existing methods developed for intensity images apply directly to finding steps in range images.

An example is shown in Figure 9.4. In this image, most of the edges detected correspond to surface boundaries, and the orientations of the edge chains are reliable indicators of the orientations of the surface boundaries. The only exceptions are immediately to the left and right of the pallet where there are regions in which there is a gradual transition from valid data on the floor closer to the sensor to no valid data further away. For polyhedral objects, the true occluding boundaries will be straight, and so the goodness-of-fit of a straight line to an edge chain can provide a good indication of whether or not the edge corresponds to an occluding boundary.

Creases

Creases are second order features; i.e. discontinuities in the first derivative of the image surface. They mark the join between two adjacent surfaces. Two main approaches have been suggested for crease extraction from range data (most other techniques are variations on these):

- finding local maxima of curvature in the image surface (Brady *et al.* [29]);
- finding local minima in the dot products of the surface normals (Faugeras and Hebert [65]).

The first of these requires the computation of second derivative operators which, it is well known, are noise sensitive. The second involves finding significant changes

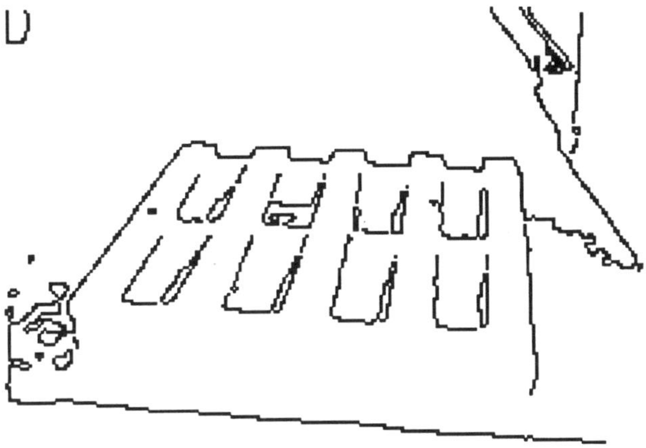

Figure 9.4: A step edge map of an NEL range image of a pallet: Canny [48] edges extracted from an NEL range image of a pallet. Most of these correspond to occluding boundaries, with the exceptions being immediately to the left and right of the pallet, where there is a gradually transition from regions with valid data to regions with no valid data (see Figure 9.3: as the angle of incidence of the laser stripe on the floor becomes more acute, there is correspondingly less Lambertian reflection towards the sensor)

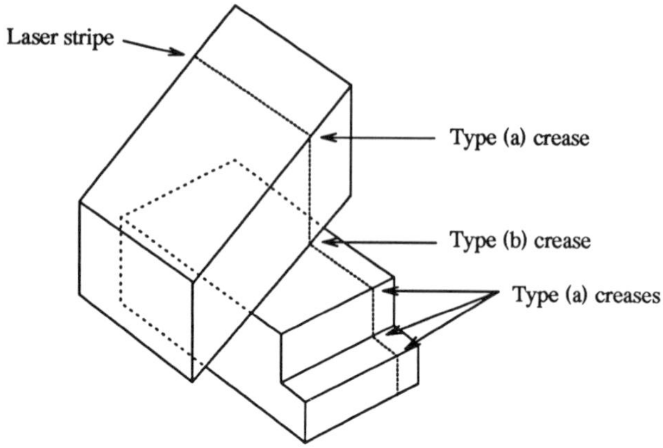

Figure 9.5: Interpretations of creases in range data: the figure depicts a laser stripe falling on two objects. Although the crease types (a) and (b) have different world interpretations (explained in the text), they are indistinguishable in the range image without extra knowledge such as object models

in the orientation of two proximate surface patches; this is exactly the same test as for point/patch merges, the Haralick F-test, mentioned previously.

A crease in range data admits two interpretations with respect to the objects in a scene. These are shown in Figure 9.5. Firstly (type (a) in Figure 9.5), it may correspond to two adjacent surfaces of the same object, in which case the patches either side of the crease embody all the matching power available (since the crease location is implicit in the plane equations of the two patches), save the knowledge that the two sensed patches should be matched to adjacent model surfaces. Secondly (type (b) in Figure 9.5), it may be the result of two abutting surfaces from different objects. In this case the crease has the same effect as an occluding boundary, namely, it constrains the extent of at least one of the two surfaces. These two cases cannot be distinguished in the range image without additional knowledge such as object models.

Apart from the problems of noise sensitivity an additional difficulty with crease extraction arises in NEL range maps because the vergence angle between the laser stripe and the camera is small, meaning that orientational differences in image space are much less than those in the scene. An example of this is given in Figure 9.6, which shows a column of a range image which resulted from viewing a right-angled crease in the scene. As discussed above with reference to patches, image surface orientation differences are used to delimit region growth, but the smallness of these differences means that accurate localization is difficult.

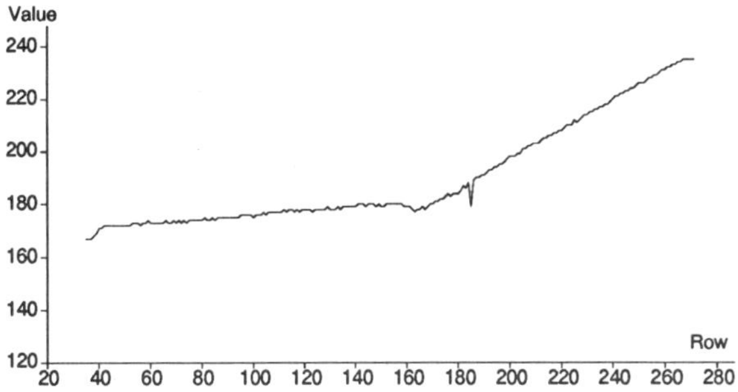

Figure 9.6: The range image profile of a right-angled crease in the scene. The value is plotted for each row in a given column

9.2 The Utility of Features for Recognition

Position/surface normal data are simple to extract, as shown in the previous section. Moreover, they are conveniently represented by two vectors, $s = (\vec{p}, \vec{n})$, and under this representation we can define viewpoint-invariant measurements on pairs of these features; constraints on such measurements have been shown to be efficient in constraining the search involved with recognition (e.g. Grimson [81], I. D. Reid and Brady [171, 175]). One such set of measurements which embodies all the possible constraints on position and orientation available from two position/surface normal data points is given by:

surf-angle $\vec{n}_1.\vec{n}_2$

surf-dist-1 $\vec{n}_1.(\vec{p}_1 - \vec{p}_2)$

surf-dist-2 $\vec{n}_2.(\vec{p}_2 - \vec{p}_1)$

surf-cross-dist $\frac{\vec{n}_1 \wedge \vec{n}_2}{|\vec{n}_1 \wedge \vec{n}_2|}.(\vec{p}_1 - \vec{p}_2)$

In this section we consider the utility of position/surface normal points extracted from range images in constraining object parameter values and compare this with the utility of higher order features: patches, steps and creases.

We begin with a recognition example based on the method described by Reid and Brady [171, 172, 175, 173] for locating instances of object classes with multiple internal parameters. In this system recognition is structured as a search of an interpretation tree using pairwise consistency of observed features to model feature matches to prune the search space. Internal parameters are represented in a table in which upper and

```
interpret(I, P, j)
[ if j = n
    [ verify(I, P);
    [ return;
  i ← 0;
  while i < m
    [ if consistent(I, P, {s_j, μ_i})
        [ P' ← update(I, {s_j, μ_i}, P)
        [ if P' ≠ ∅
        [   [ I ← I ∪ {s_j, μ_i}
        [   [ interpret(I, P', j + 1)
    [ i ← i + 1
  return;
```

Figure 9.7: Pseudo-code for the interpretation tree search for a match from n observed data features, $s_0 \dots s_{n-1}$, to a model with m features, $\mu_0 \dots \mu_{m-1}$. The symbol I represents the current interpretation, that is, a list of pairwise consistent data to model feature matches. The current network state P (i.e. estimated upper and lower bounds on the object's internal parameters) describes a multi-dimensional rectangular space. The function `consistent` returns true if a given data to model feature match is pairwise consistent with all other matches in the current interpretation. The function `update` triggers network propagation which terminates either when it has converged to a steady state, or the upper and lower bounds of a network variable crossed during network evaluation, flagged by the condition $P' = \emptyset$

lower bound estimates (suprema and infima) for each parameter are stored, and the set of functions which relates the internal parameters to observable measurements, and the relationships between parameters, is represented by a real-valued constraint satisfaction network of the type described by Fisher and Orr [67]. Figure 9.7 shows a pseudo-code version of the search algorithm.

Recognition examples

Figure 9.8 shows two objects and parameterizations for these. The object represented by (a) is referred to as the "widget" for convenience, and is parameterized by the set of size parameters (defined in Figure 9.8):

$$\{w_1, w_2, w_3, h_1, h_2, h_3, d_1, d_2, d_3, d_4, d_5, d_6\}$$

The object in (b) is referred to as the chimney for convenience, and is parameterized by the set of two angle and fourteen size parameters:

$$\{\theta, \alpha, w_1, w_2, w_3, w_4, w_5, w_6, h_1, h_2, h_3, h_4, d_1, d_2, d_3, d_4\}$$

Synthetic range images of instances of these two classes are shown in Figures 9.9(a) and (b). A number of position/surface normal data estimated from the range

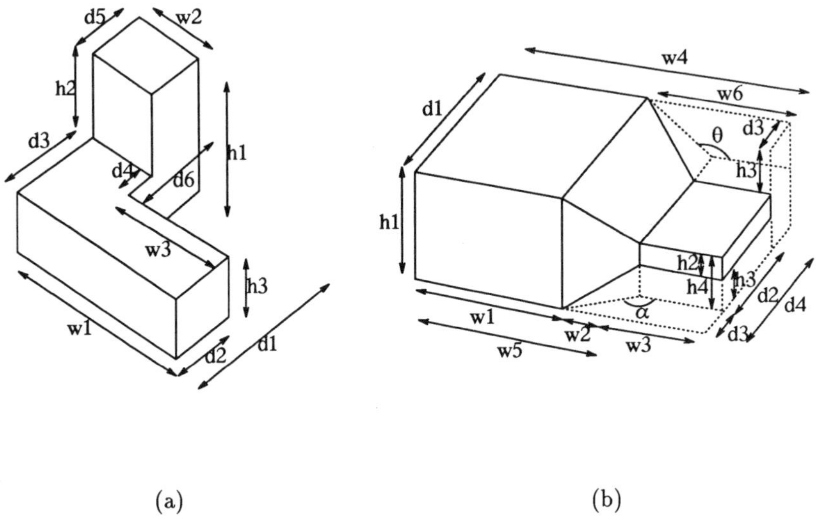

Figure 9.8: Parameterizations for two objects: (a) a widget; (b) a chimney

data are indicated by small vectors and a point number, i, for observation $(\vec{p}_i, \vec{n}_i)$ (the position and normal respectively). Uncertainty in these points is only due to slight aliasing effects, which means that uncertainty in the computed parameter ranges is attributable to a lack of evidence rather than to large error estimates.

Legal interpretations for these point data and models, and valid parameter ranges were derived using the constrained search and network propagation algorithm (see Figure 9.7). Wireframes for the class instances are shown superimposed over the range images in the computed poses. The parameter ranges for the interpretations shown are given in Figures 9.9(c) and (d)[1], however the ranges for the two angle parameters of the chimney are not shown in that figure since it would be meaningless to compare them against distances.

Their ranges (in radians) were computed to be

$$\theta \in [2.58, 2.65] \quad (\text{true value} = \tfrac{5\pi}{6} \approx 2.62)$$
$$\alpha \in [2.26, 2.46] \quad (\text{true value} = \tfrac{3\pi}{4} \approx 2.36)$$

Note that the wireframe is drawn to satisfy the low bounds of the parameters (since high bounds are not always available).

[1] An arrow indicates that no upper bound was found.

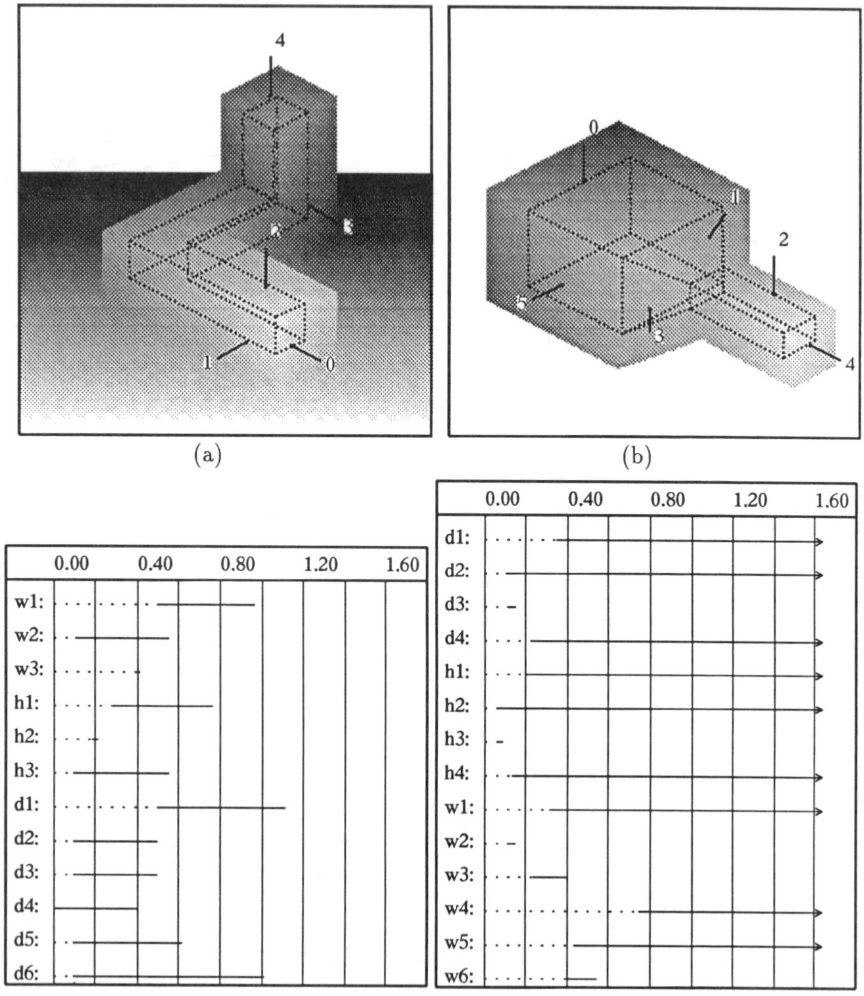

Figure 9.9: Recognition examples in synthetic range data: (a) and (b) show synthetic range images of the widget and chimney, respectively (lighter is closer, white is no data). Points from the object surfaces have been selected manually (and labelled 0,1 etc), and the surface normals estimated from the range data in a neighbourhood of each point. In each image, a wireframe, computed by instantiating the boundary representation with the lower bound of the parameter values, is shown superimposed in the computed pose. (c) and (d) show the computed parameter ranges for the widget and chimney, respectively

We now use these examples to examine the effectiveness of the point (position/surface normal) data representation in determining the parameters of the objects.

The utility of points

Firstly, we note that in some cases, very good bounds have been computed (h_2 and w_3 for the widget, d_3, h_3 and w_2 for the chimney). This is because in each of these cases the point data contains all the information required to determine both upper and lower bounds on the parameter. For example the image measurement surf-dist leads to the feature constraint on w_3:

$$w_3 \leq \vec{n}_3.(\vec{p}_0 - \vec{p}_3) \leq w_3$$

and the final computed range for w_3 was

$$w_3 \in [0.39, 0.41] \qquad \text{(true value} = 0.4)$$

The next conclusion to draw is that, for many of the other parameters, the point estimates from Figures 9.9(a) and (b) have led to poor lower bound estimates. The chimney parameter, d_1, was affected by the feature constraint surf-dist:

$$-d_1 \leq \min_{j=0,1} \vec{n}_5.(\vec{p}_j - \vec{p}_5) \leq 0$$

and the final computed range for d_1 was

$$d_1 \in [0.35, \infty] \qquad \text{(true value} = 0.45)$$

In these range images, there is clearly much more constraint available for the parameter d_1 and therefore neither of the points 0 nor 1 is a particularly good representation of the top surfaces of the object.

The utility of patches

Consider the block shown in Figure 9.10. The feature constraint for a point in patch P on the top face, projected onto the front face, the plane defined by $(\vec{p}, \vec{n})$, is

$$-w \leq \min_{\vec{x} \in P} \vec{n}.(\vec{x} - \vec{p}) \leq 0$$

and any point of P lying on the far edge of the block provides the maximum possible constraint on the lower bound of w. Thus, patches can provide tighter constraints on lower bounds than those afforded by arbitrary points. In addition a patch representation leads to many fewer data to match, giving a shallower search. Note, however that without the additional knowledge that we have obtained the full extent of the patch, no further constraint on the upper bound is possible.

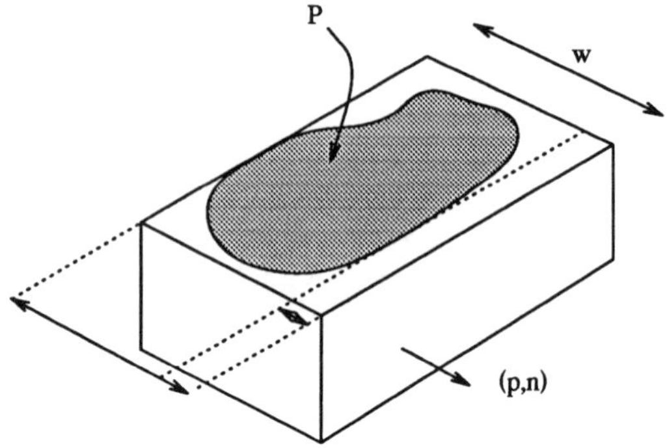

Figure 9.10: The utility of a patch versus a point

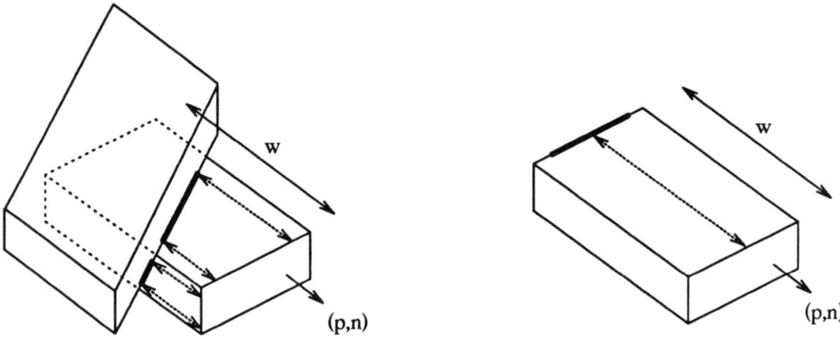

Figure 9.11: Parameter bound estimates from edges

The utility of step and crease edges

We now address the utility of edge locations. We remarked above that points on the far edge of Figure 9.10 give the best possible lower bound estimate. Similarly, in Figure 9.11(a), the best lower bound estimate is given by the furthest of the crease segment's endpoints and the occluding edge's endpoints from the front face. Additionally, (b) illustrates that in some cases edges provide information about the upper bounds on the size parameters; in this example the location of the edge marked in bold, relative to the surface defined by $(\vec{p}, \vec{n})$, bounds the value of w above as well as below.

General observations about the various edge types are:

- the near side of an step edge with valid data on both sides definitely delimits a surface;

- if one side of a step does not have valid data then it may or may not be the boundary of a surface;[2]
- creases may or may not delimit a surface, and in any case are difficult to find;
- the far sides of occluding boundaries probably do not delimit surfaces.

9.3 Feature Extraction and Matching Strategy

In the preceding sections we considered how features can be extracted from a range image, and how effective various features are in determining the unknown model parameters of a class instance. The point-based data representation leads to simple, elegant feature constraints, given in Section 9.2, but does not afford the same constraining power of the higher level features; patches and edges.

We now show how simple extensions to the point representation, and to the interpretation tree search, can harness much of the additional constraint provided by higher level features. In this way we gain the simplicity of the point-based feature constraints (and do not need extra feature constraint tables for comparing edges with edges and/or edges with point/patches) without compromising too greatly the ability to constrain the parameter values as much as possible.

Firstly, we observe that points on the boundary of a patch can be used to represent the patch. In particular, usually a few key points (i.e. position/surface normal data) delimit the patch entirely. These points are matched as before to the model with the additional matching constraint that they all match the same model surface. Figures 9.12(a) and (b) show the same objects and data as in Figures 9.9(a) and (b), respectively, but with additional points taken from close to the boundaries of the surfaces. The improvement in the lower bound estimates is obvious from the improved fit of the wireframe to the range data.

Now suppose that points 5, 6 and 7 on the widget, and points 6, 7, 8 and 9 on the chimney are known to lie on occluding boundaries; i.e. they lie on step edges forming parts of the patch boundaries. This means that they lie on *two* model surfaces, albeit one of them invisible. Having established the consistency of a match for one of these points to a model surface relative to existing assignments in the current interpretation, a secondary match for the same point is sought from amongst the adjacent faces in the model. Because the surface is invisible no sensed surface normal is available, however one of the two surf-dist image measurements can still be determined, and it is this one which gives the upper bound constraint. Figure 9.13 illustrates secondary matching for a point $(\vec{p}_2, \vec{n}_2)$ which is known to lie on both the right face and the back face of the block. The search assigns point $(\vec{p}_2, \vec{n}_2)$ to a model face μ_j and then checks the consistency of a second match to an adjoining surface based on $\vec{p}_2$ alone.

[2]Often this can be resolved by how far away the surface is and how straight the boundary is.

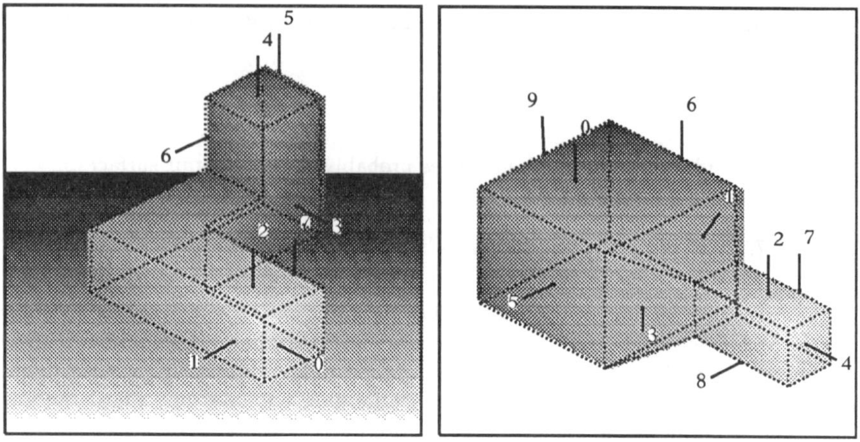

(a) (b)

Figure 9.12: Wireframes computed using improved parameter estimates: the addition of point data on and/or near to surface boundaries of the objects to the scenarios depicted in Figure 9.9, gives a computed sizes of the instantiated boundary representations much closer to the true sizes of the objects in the scene

The correct match leads to the feature constraint

$$-w \leq \vec{n}_1.(\vec{p}_2 - \vec{p}_1) \leq -w$$

Two further advantages of secondary matching to invisible faces exist:

- The orientation of the invisible face was not sensed, and therefore is initially unknown, meaning that only one feature constraint can be used. However, having once estimated the rotation of the object, the unknown surface normal can be replaced by the rotated model surface normal, meaning that all four feature constraints involving the invisible face become available. This can lead to dramatic improvements in the parameter estimates (see Figure 9.14).

- Secondly, unlike a point matched in the interior of a surface which constrains only one translational degree of freedom, a point matched on an occluding boundary constrains two translational degrees of freedom, providing more constraint for the estimation of the translational component of the model to sensor transformation.

Figure 9.14 shows bars representing the ranges of parameters computed using three different point-based representations. Three bars for each parameter are given, one for each of the results using (from top to bottom, respectively):

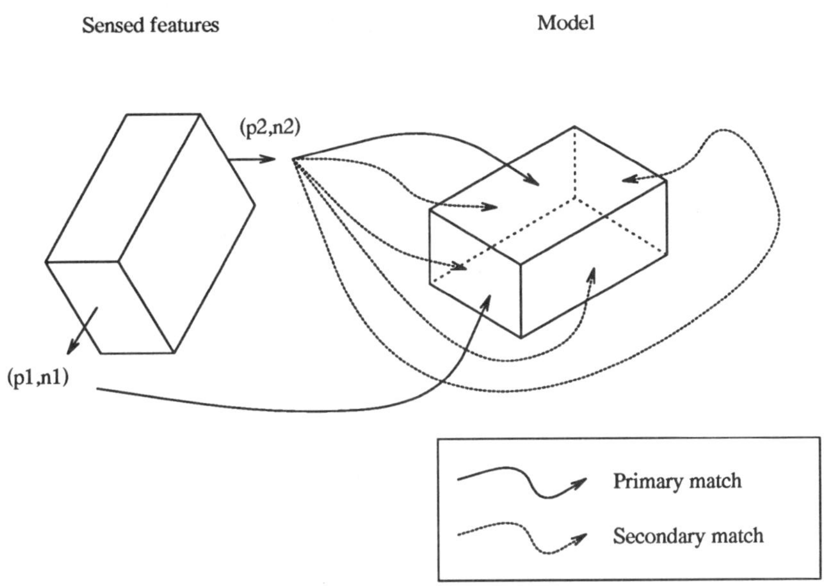

Figure 9.13: Secondary matching: given a pair of primary matches $s_1 = (\vec{p}_1, \vec{n}_1) \rightarrow \mu_i$ and $s_2 = (\vec{p}_2, \vec{n}_2) \rightarrow \mu_j$, a secondary match for s_2 can be sought from among the surfaces which share a boundary with μ_j

- point data in the interior of patches;
- point data from the boundaries of patches;
- point data from boundaries with secondary matching.

The dramatic improvement in both upper and lower bounds when using secondary matching is clear.

If secondary matching of a feature on the near side of a true occluding boundary (one with valid range data on either side of the edge) fails, then the search backtracks, since this range image feature admits only one interpretation, that of surface boundary. For the other cases secondary matches are attempted but need not be successful. Failure in these situations means that there is no consistent way of interpreting the point as lying on two model surfaces. This may be correct and therefore the search continues without backtracking.

Finally, we describe one additional extension. The secondary matching method described above does not use any edge orientation information which could further restrict searching and improve the efficiency of the bounds computation. The angles between an edge (when its orientation is known) and the sensed normals are checked for consistency with the model values. Then, given the angle, θ, between the primary surface and the invisible one (available from the feature constraint tables), the primary surface normal, $\vec{n}$, and the edge orientation, $\vec{e}$ (both are sensed values), it is straightforward to compute the surface normal of the invisible face. This is given by

$$(\cos\theta)\vec{n} \pm |\sin\theta|(\vec{n} \wedge \vec{e})$$

where $\pm$ indicates that the direction of $\vec{n} \wedge \vec{e}$ should be adjusted so that it points away from the viewer. This estimate of the invisible surface orientation subsequently allows the use of all four feature constraints. If θ is not known exactly (it could be a parameterized quantity) then the range of values is treated analogously to estimated errors in sensed orientations.

9.3.1 Pallet Recognition

The examples above illustrate the principles using synthetic data. We conclude with an example which brings together the feature extraction methods outlined in Section 9.1 with the extended recognition strategy given in this section to achieve one of the major aims of the AGV project, that of pallet localization. Figure 9.15 depicts a simplified model of a pallet parameterized by the set:

$$\{w, h, d, s, w_c, h_c, h_t, g\}$$

Figure 9.16 shows an scene containing two substantially occluded pallets (a), and Oxford/NEL range image of the scene (b). Figure 9.17 gives planar surfaces extracted (a) and (b), the computed size and pose of the pallets (c) and (d), and the computed parameter ranges (e) and (f).

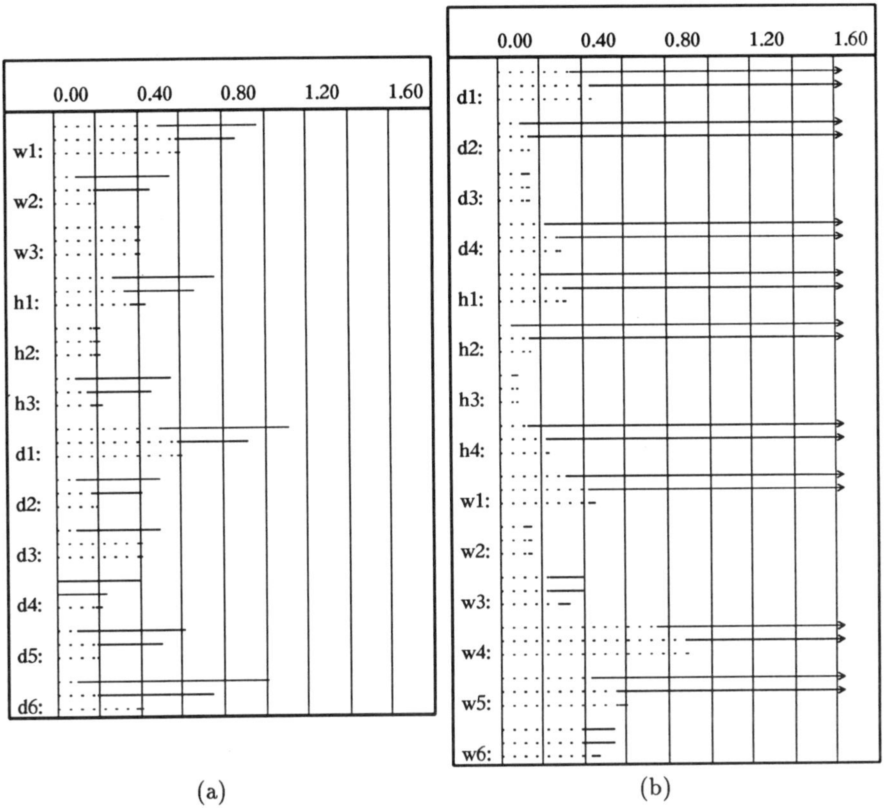

Figure 9.14: Using secondary matching to improve the parameter bounds: the solid bars next to each parameter represent the computed ranges for different experiments. (a) shows the parameter ranges for the point data of Figure 9.12(a), from top to bottom: matches for points 0–4 only; matches for all points; and matches for all points, with secondary matches for points 5–7. (b) shows the parameter ranges for the point data of Figure 9.12(b), from top to bottom: matches for points 0–5 only; matches for all points; and matches for all points, with secondary matches for points 6–9

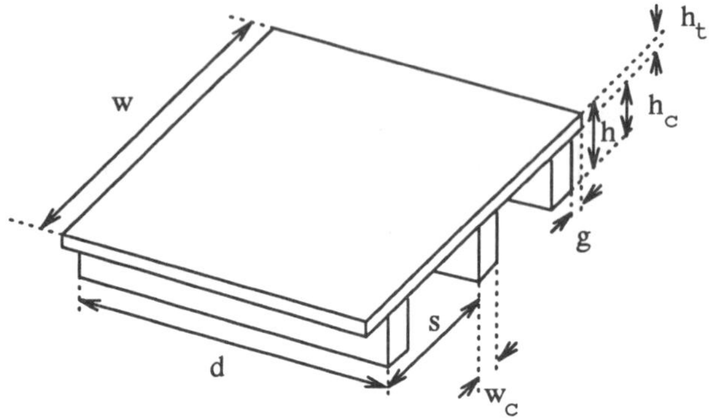

Figure 9.15: The simplified pallet model: the model contains a top surface, parameterized by $\{w, h_t, d\}$, and a set of struts parameterized by $\{w_c, h_c, d\}$. The strut positions relative to the top surface are governed by the overlap at each end, g, and the strut separation, s

(a) (b)

Figure 9.16: A scene containing two pallets and a cardboard widget: (a) an intensity image of the scene; (b) an NEL range image of a similar scene in which lighter represents closer, and white means no valid range data

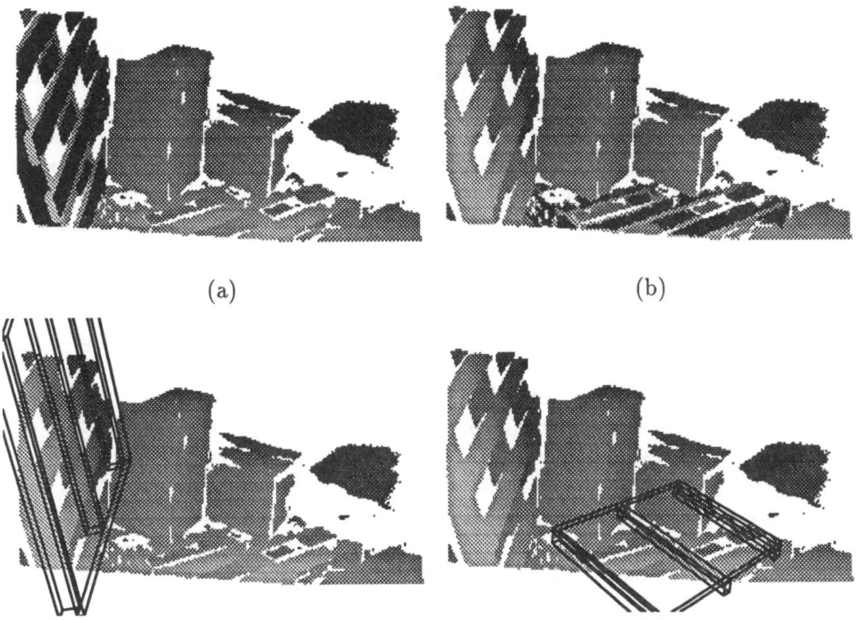

(a) (b) (c) (d)

	computed		measured
	inf	sup	(by hand)
w_c	19	76	40
h_c	68	381	85
h_t	0	67	20
w	765	1143	885
d	502	∞	885
s	334	381	385

(e)

	computed		measured
	inf	sup	(by hand)
w_c	13	38	30
h_c	45	188	90
h_t	0	66	15
w	685	900	810
d	334	∞	1060
s	336	419	390

(f)

Figure 9.17: Pallet recognition: (a) the surfaces extracted from the upright pallet; (b) the surfaces extracted from the prone pallet; (c) the legal interpretation for the upright pallet; (d) the legal interpretation for the prone pallet; (e,f) the computed parameter ranges in mm for the interpretations in (c) and (d), respectively

9.4 Summary

In this chapter we have advocated basing feature extraction on reliably finding planar patches, and then representing them as point features. The feature extraction algorithm detects steps using a standard intensity edge detection method (Canny [48]) and then finds planar patches by clustering neighbouring points which have the same (or nearly the same) surface normal, and which are not separated by step edges. A small number of points on the boundaries of a patch contain most of the information about the extent of the patch and therefore these are used to represent it.

Pairs of point features give rise to simple geometric measurements which are well suited to pruning the interpretation tree, but they exhibit deficiencies in determining object parameter bounds, especially upper bounds. In order to overcome this problem we have devised extensions to the basic tree search algorithm which take advantage of knowledge that a point lies on an edge. In the examples shown, only a few points on the object boundaries, in combination with the extended search, were sufficient to obtain tight bounds on all object parameters.

The underlying philosophy of this chapter is one of simple, local features, extracted using simple but strategic algorithms so as to optimise the constraint on parameter bounds. This philosophy, however, differs from many previous researchers, who have used complex image segmentation algorithms (e.g. Blake and Zisserman [19], Silverman and Cooper [182], Yokoya and Levine [197]) and feature grouping techniques (Fisher [66], Dickson [58]) as precursors to constrained search. Many "state-of-the-art" segmentation algorithms are slow and unreliable, moreover, we have demonstrated in this chapter that such techniques are not necessary for efficient recognition.

Part III

MODEL BASED PLANNING

Chapter 10

Introduction to Path Planning

Stephen Cameron

Any autonomous vehicle must be able to move around without colliding with obstacles in its environment, and this requirement makes path planning a basic skill with which all guided vehicles must be endowed with. However it is only recently that the basic principles for path planning have emerged in a coherent form. Path planning is difficult, and this difficulty is not just an artifact of the problems of detecting where the obstacles are; there are also fundamental problems in characterising the allowable regions that the robot can use without a collision occuring. To most people this may seem strange, as humans can normally manage to avoid bumping into things without much in the way of conscious effort. The situation may be compared with vision, which humans do far better than computers despite decades of research: we cannot just take a 'core dump' of a human brain to determine how it's done!

Early work on robotic path planning concentrated on robot manipulators, mainly because it was a practical problem that needed solving. Rather than just giving a history of the solution of the path planning problem we will build up the techniques gradually, and concentrating on those that are particularly relevant to solving the path planning problem for vehicle systems.

10.1 The Railway Track Algorithm

A high-speed railway system that relied on the eyesight of its drivers to keep the trains apart would be doomed to failure. However the fact that the trains are constrained to travel along tracks makes the path-planning problem quite simple—at least for one vehicle. Namely, we can treat the problem as one of traversing a mathematical graph structure, in which the graph edges represent tracks and the graph nodes represent

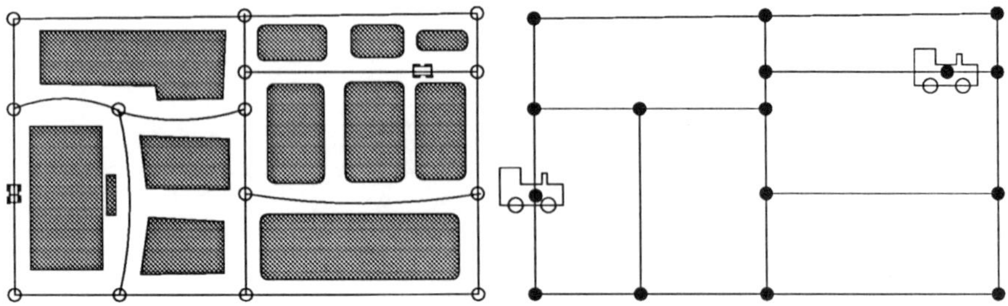

Figure 10.1: Establishing railway tracks

the joins between tracks (at points). As a simple example, the layout in Fig. 10.1(a) could be mapped into the graph of Fig. 10.1(b), in which the exact 'shape' of each 'track' is abstracted away, and we add extra nodes to mark the start and end of the path.

For finding the shortest route in this graph several algorithms are available. For example, Dijkstra's algorithm (Fig. 10.2) works by storing at each node a number that is the shortest distance found from the start node, set initially to 0 for the start node (`S`) and ∞ for ever other node. A *working set* of partial paths is maintained—these are paths which are yet to be expanded—and this set is initially set to the trivial path that consists of only the start node. At each stage we select (and remove) some path from the working set, and consider all the nodes we could get to by following that path, then following one further edge. If this would result in a path that is a shorter distance from the start node than we found before, we update the shortest distance (`dist`) field for that node, and place this new path into the working set (where `P+n` means the path `P` followed by the traversal to the node `n`). The algorithm terminates when the shortest distance from the start node to *every* node in the graph has been found. The shortest path to the goal can either be stored as it is found, or we can trace back from the goal using the values stored at the nodes.

As an example, consider Fig. 10.3(a), which shows a small graph with nodes labelled S (start), G (goal), and A–D, and the edges labelled with a length. For this example we will assume that we always **select** the shortest path in the working set, so denote a path by a pair of path nodes and length, as in $\langle SAC, 9\rangle$. Then the initial working set is $\{\langle S, 0\rangle\}$, leading to the following sequence of working sets:

(b) $\{\langle SA, 3\rangle, \langle SC, 5\rangle\}$
(c) $\{\langle SAB, 7\rangle,$ ~~$\langle SAC, 9\rangle$~~, $\langle SC, 5\rangle\}$
(d) $\{\langle SAB, 7\rangle,$ ~~$\langle SCA, 11\rangle$~~, $\langle SCD, 9\rangle\}$
(e) $\{$~~$\langle SABD, 13\rangle$~~, $\langle SABG, 10\rangle, \langle SCD, 9\rangle\}$
(f) $\{\langle SABG, 10\rangle,$ ~~$\langle SCDG, 14\rangle$~~, ~~$\langle SCDB, 15\rangle$~~$\}$
$\{$~~$\langle SABGD, 15\rangle$~~$\}$

```
forall nodes v do dist( v): = ∞ endfor;
dist( S) := 0;
Paths := {[S]};
while Paths ≠ {} do
    select P from Paths;
    t := head( P);
    foreach neighbour n of t do
        newdist := dist( t) + length( edge( n, t));
        if newdist < dist( n) then
            dist( n) := newdist;
            Paths := Paths ∪ {P+n};
        endif;
    endfor;
endwhile;
```

Figure 10.2: Dijkstra's algorithm for shortest path

where the labels refer to the snapshots shown in Fig. 10.3, and the crossed paths are briefly considered during the **foreach** loop, but do not actually make it into the working set as there is already a shorter path known. Fig. 10.3(b)–(d) also show the finite distances found to each node, the termination points of the paths in the working set (circled), and which path is to be expanded next (filled circle).

There is flexibility in the algorithm in deciding which path to **select** at each stage. We showed 'best-first' search above, in which we select the 'most promising' (i.e., shortest) path at each stage. Another strategy is to select the paths that visit the smallest number of nodes first, giving *breadth-first* search as shown in Fig. 10.4, in which the dotted lines encapsulate the 'wavefront' of the search. A third strategy is to simply do a depth-first search, by selecting the last path to enter the working set at each stage. Fig. 10.5 illustrates this, in which the current path is enhanced, and terminal nodes of paths waiting to be expanded are indicated by arrows after (a) one stage, (b) two stages and (c) five stages; after this the paths $SCDB$, $SCDBA$, SCA, SA, SAB and $SABG$ will be explored and cause the distances stored to be updated. In general, best-first search will use expand fewer paths, but requires some sort of search of the working set to find the next path to expand; depth-first search is the simplest to implement as we can use recursion and do not need to explicitly store the working set (we can use the programming language's call stack instead); and breadth-first search is a compromise between the two. However note that in all cases the entire graph will be explored.

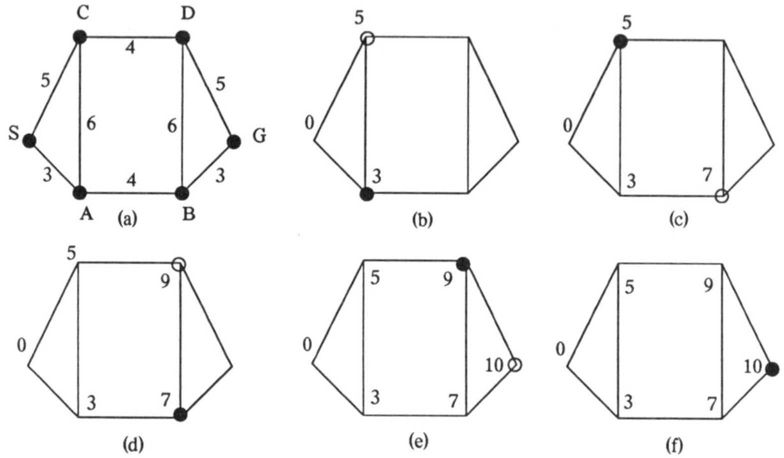

Figure 10.3: An example of Dijkstra's algorithm (best-first search)

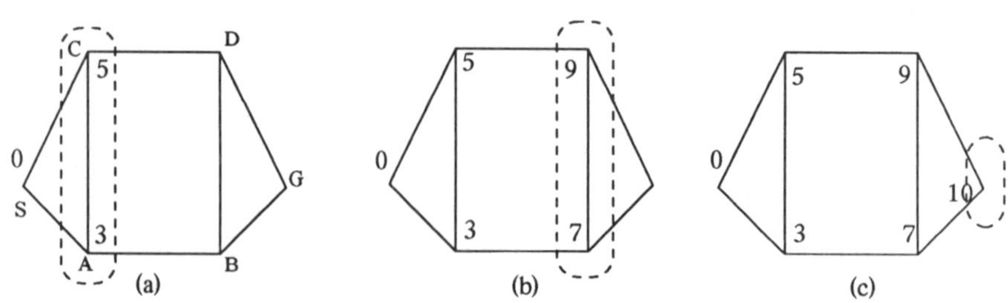

Figure 10.4: An example of Dijkstra's algorithm (breadth-first search)

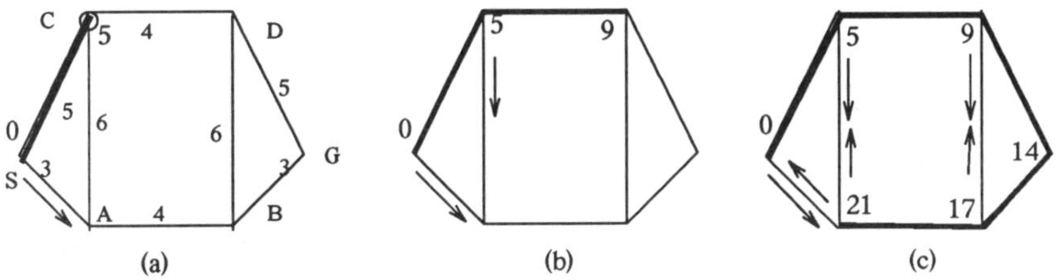

Figure 10.5: An example of Dijkstra's algorithm (depth-first search)

10.2 Moving Off the Tracks

Once we have vehicles that are free-roving then the number of routes that a vehicle can take is literally infinite, and to decide on a collision-free path we need to restrict the number of paths we consider somehow. The trick here is to consider enough paths of the right sort so that we don't miss a short path that does exist[1]. So the first problem is to characterise the possible paths.

One way, that works if the vehicle (or vehicles) have a lot of space to manœuvre, is to essentially lay down 'virtual railway tracks' through each 'corridor' in the environment; then, despite the fact that any vehicle may not lie on the tracks all the time, we can imagine that we *could* move a vehicle to the nearest track at the beginning of a path, move to somewhere near the goal position, then move off the track at the end of the path to that goal position. In this case, each piece of 'track' characterises a portion of the space of all possible paths, *viz.* the paths that use a particular piece of a natural corridor. In many cases, however, there is not enough room to manœuvre for this approach to work without added refinement.

10.3 Configuration Space

When we have to take the shape of the robot into account path planning is complicated by the fact that all parts of the robot must simultaneously avoid collisions with the environment. Configuration space is a notion that was formalised and popularised by Lozano-Pérez that enables us to separate out worries of the form 'which parts of the objects might collide' from worries about characterising potential paths.

Although configuration space (or *C-space*) is a concept that can in theory be applied to any type of robot, it will be best to consider only simple examples first. The environment for most vehicular robots can be considered to lie in a plane, and

[1]Athough if the time that we can take to plan the paths is limited (as it often is) we may loosen this requirement a little.

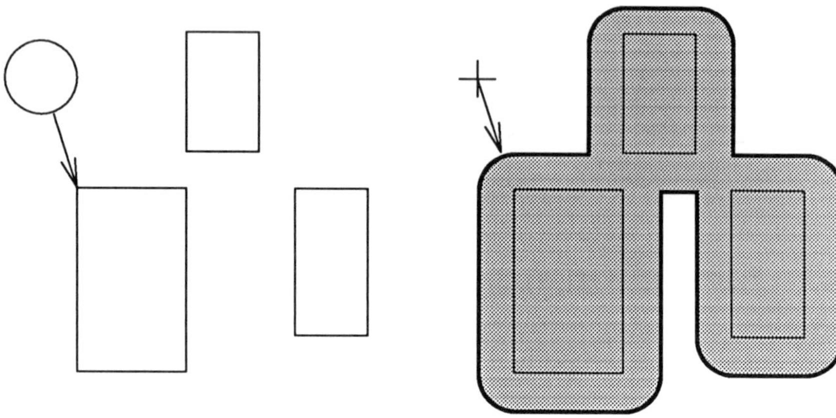

Figure 10.6: C-space for a cylindrical robot

some vehicles have an approximately circular projection onto that plane. For these *cylindrical robots* we can replace the problem of finding a path that keeps all parts of the robot away from obstacles into the problem of keeping the *centre* of the robot a distance greater than the radius of the cylinder away from the obstacles. Thus we can mark out a new map of the environment in which we 'shade in' the space that lies too close to the obstacles; finding a collision-free path in this new map for the centre of the cylinder yields a collision-free path for the robot also.

As an example, Fig. 10.6 shows both a cylindrical robot and its workspace, together with its 'new map'; the latter is precisely the C-space for this problem, in which for any possible intermediate position of the robot we can tell immediately whether it causes a collision from whether that position (characterised here by the position of the centre of the cylinder) lies within a 'shaded' region (or *C-space obstacle*). (This map of C-space with translations only has another interesting property: the shortest translation of the robot for it to touch the obstacles is the same as the shortest distance from the origin of C-space to the C-space obstacles, as shown by the arrows in Fig. 10.6 [46].)

We can apply the same mathematical trick to other types of robot (including non-vehicular robots). For example, if we now model the robot as an arbitrary shape, that can translate in any direction but cannot change its orientation, then the set of possible intermediate positions for the vehicle is simply the set of all possible translations, and the C-space obstacles are those translations which would cause the

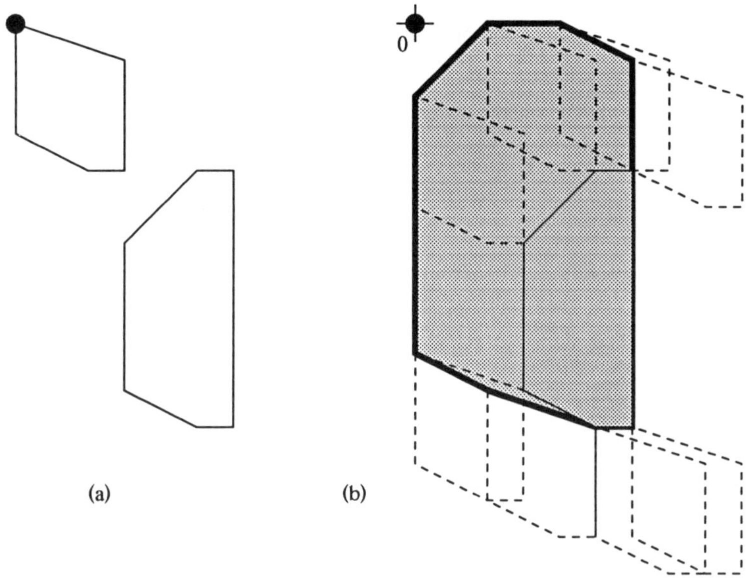

Figure 10.7: The sliding algorithm

robot to collide with some parts of its environment.

Of course we cannot consider all of the infinite number of possible translations separately in order to compute the C-space obstacles, but there is a standard method for computing the C-space obstacles when the vehicle and the environment are composed of convex polygons. Each pair of convex polygons (i.e., the robot and a part of the environment) has a corresponding C-space obstacle, and this can be determined by first translating the robot until it just touches the other polygon, and then tracing out the translations of the robot as it is slid around the polygon. (For example, the highlighted point in Fig. 10.7(a) traces out the shaded polygon in Fig. 10.7(b).) This produces a C-space obstacle which is another convex polygon, and as the polygons are convex we can determine the extent of each side of the C-space obstacle by local tests, which results in a linear-time algorithm to generate it. The sliding algorithm also easily generalises for shapes other than convex polygons, but the details and time complexity are considerably more difficult. As a result almost all implementations insist on shapes that can be decomposed into the union of a set of convex pieces.

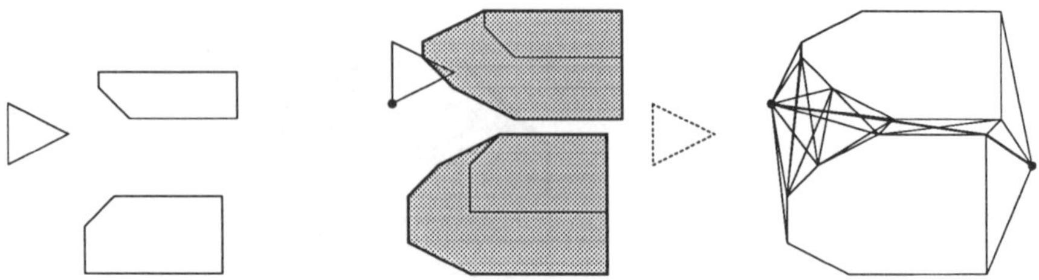

Figure 10.8: Constructing a visibility graph

10.4 Searching C-space

Once we have constructed the C-space obstacles (and stored them using data-structures similar to those used in CAD systems) we are ready to search the obstacle-free parts of C-space (or *free-space*) for a path. As before, we first have to choose a set of characteristic pieces of paths, and once this is done we invariably reduce the problem to that of searching a mathematical graph structure.

10.4.1 Visibility Graphs

The simplest way of forming a graph in the translation-only case is to use the vertices of the C-space obstacles, plus two points corresponding to the start and goal positions, as the nodes of the graph, and to join two nodes by an edge if and only if we can get from one position to the other by a single translation motion—or equivalently, if the line between the two points in C-space does not pass through any C-space obstacles. As an example, Fig. 10.8(a) shows a triangular robot within a polygonal environment, Fig. 10.8(b) shows its configuration space, together with a goal position, and Fig. 10.8(c) shows the corresponding graph, called a *visibility graph*, together with the shortest path found in that graph. As well as being simple to implement this method always finds the shortest-length path. However these paths have the distinct disadvantage (at least for robot vehicle work) that they cause the vehicle to graze obstacles as they pass, and as vehicles are difficult to control with any precision this will result in real collisions between the robot and its environment.

10.4.2 A Search*

For real problems the mathematical graphs generated can get quite complex. For example, the problem in Fig. 10.9(a) generates the configuration space of Fig. 10.9(b), and thus the visiblity graph of Fig. 10.9(c); all of this graph will have to be carefully explored if Dijkstra's method is used, including parts that are a long way away from the shortest route. However by using A* search instead a significantly smaller part of

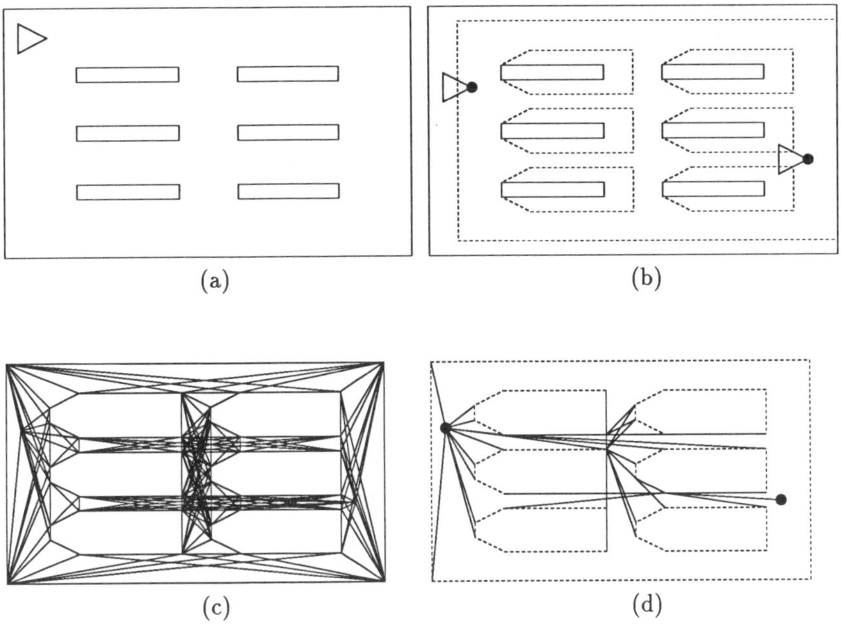

Figure 10.9: A* search

the graph need be explored, as shown in Fig. 10.9(d). Whereas Dijkstra's method just uses information about the upper bound of the length of the shortest path through all the nodes, A* works by computing a lower bound as well (Fig. 10.10). Given any intemediate node N that is being explored between the start node S and the goal node G, as we know the coordinates of the N in G (in C-space) we can work out the shortest possible distance to go between N and G, namely the straight-line path without taking into account any obstacles. We can also compute the shortest possible distance from S to N (or look up the shortest distance if we have already determined it). Thus for any potential path 'from S to G via N' we have a number that is the best possible distance that we might achieve. By keeping a set of such partially explored paths, and always expanding the most promising path first, we may finally reach a stage where we have found a true path of length l, and all of the potential paths that are left have length $> l$, at which point we can exit. (A* was also used in Chapter 4.)

In Fig. 10.11 we can compare the effect of A* with Dijkstra'a algorithm (Fig. 10.3). Fig. 10.11(a) shows the nodes of the graph, but this time also marking the lower bound distances to the goal (to one decimal place). We can denote the partial paths by a triple whose third component is the lower-bound on any path to the goal

```
forall nodes v do dist( v): = ∞ endfor;
dist( s) := 0;
shortest := ∞;
Paths := {[s]};
while Paths ≠ {} do
    select P from Paths with minimum minlength( P);
    t := head( P);
    foreach neighbour n of t do
        newdist := dist( t) + length( edge( n, t));
        if newdist < dist( n) then
            dist( n) := newdist;
            Paths := Paths ∪ {P+n};
        endif;
    endfor;
endwhile;
```

Figure 10.10: A* algorithm for shortest path

that starts with this partial path. Thus we obtain the following snapshots of the working set:

(b) $\{\langle SA, 3, 9.3\rangle, \langle SC, 5, 12.2\rangle\}$
(c) $\{\langle SAB, 7, 9.8\rangle, \langle SC, 5, 12.2\rangle\}$
(d) $\{\langle SABG, 10, 10.0\rangle, \langle SABD, 13, 17.7\rangle, \langle SC, 5, 12.2\rangle\}$

No further expansions are required, as all the other possible paths (that start with elements from the working set) must be longer. Although this is an extreme example, A* is reasonably simple to implement and efficient enough to be a popular choice of search mechanism. More complex mechanisms are possible: for example, by searching simultaneously from both the start and the goal.

The added complexity of A* makes Dijkstra's method a reasonable choice for small numbers of nodes (say, < 50), but A* tends to be an increasing better choice when there are large numbers of nodes that are far away from the direct path. Not that both methods can be used when the optimality criterion is not just total distance, providing that for A* we can still generate a lower bound on the distance still to go. Other useful measures include energy expenditure and time.

As we noted above these graphs can get quite complex. However it is often possible to examine a graph *lazily*—that is, only generate as much of the graph as is really required. For example, we can generate a visibility graph lazily by starting only with the nodes, and only generating edges (which involves the relatively expensive visibility test) as they are required in the **foreach** loop.

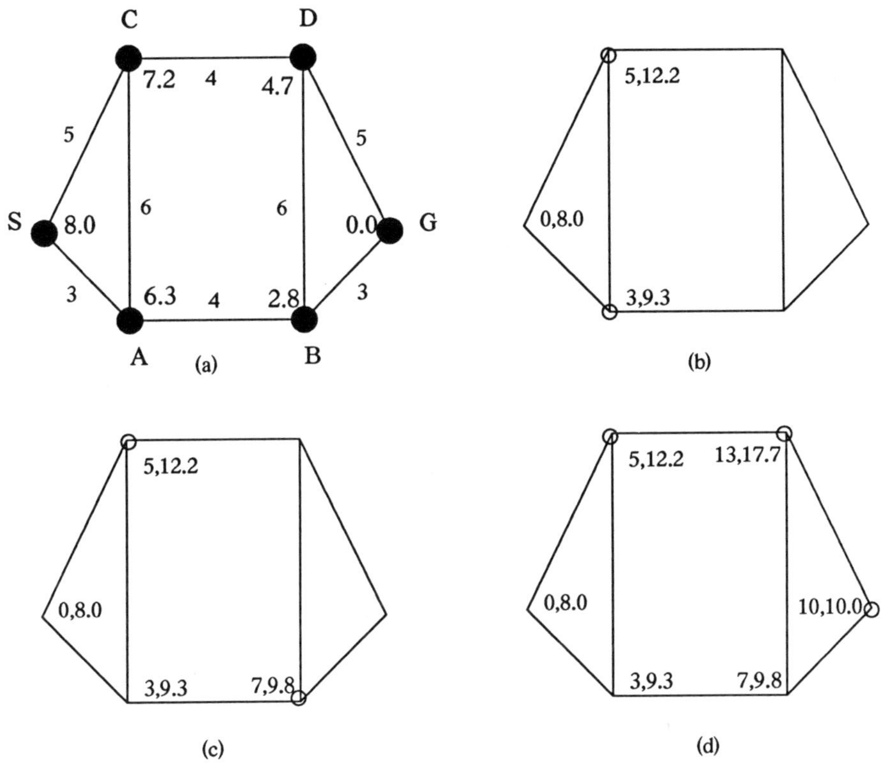

Figure 10.11: A detailed example using A*

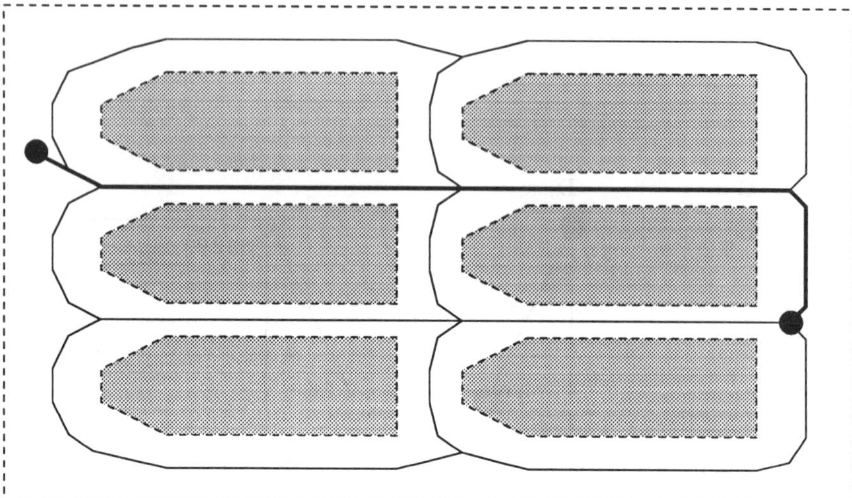

Figure 10.12: Voronoi diagram

10.4.3 Getting Away from it All

The problem of controlling a vehicle to just miss corners may tempt one to try to keep the vehicle as far away from the obstacles as possible. A straightforward way of doing this is to use a geometric data-structure called the *Voronoi diagram*. The simplest form of Voronoi diagram is defined for discrete sets of points, when it is the set of points that are equidistant from two (or more) of the points in the input set. Such a diagram forms a graph of straight-line segments, and is easily computed. For our purposes the generalised Voronoi diagram (GVD) is more useful, and is the set of points that are equidistant from two or more *shapes*. With convex polygons as our shapes the diagram is composed of straight and parabolic line segments; computationally efficient but otherwise very complex methods are known for computing it, as well as a number of slightly less-efficient but more comphrehensible methods. These line segments provide a basic set of 'tracks' for us to navigate around the free-space, and we also have to design motions to move the vehicle onto a track near the start, and off a track near the goal. Fig. 10.12 shows an approximation to such a GVD, together with an example path.

We can view the Voronoi diagram technique as being a way of providing a set of 'roadways', which we can use rather like the railway tracks of §10.1 but with an added requirement to navigate onto and off of the tracks at the start and the goal. There is no general method that always provides the 'best' roadways, as the notion of optimality will often take into account factors which are not strictly geometric. This point is taken up again in Chapter 11 and Chapter 13.

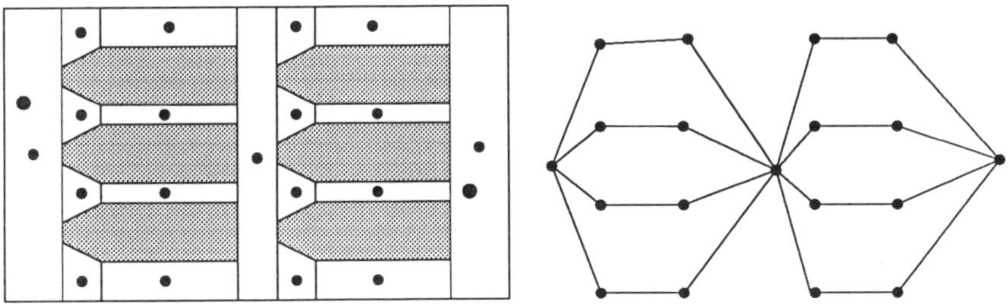

(a) (b)

Figure 10.13: Region decomposition

10.4.4 Region Decomposition

If we view the Voronoi diagram technique as a way of generating lines in the road for a vehicle to follow, the region decomposition technique generates just a *list* of roads instead. The idea here is to take the *free space* for the robot (i.e., the space outside obstacles) and to partition it up into regions (the 'roads'). Then we can build a connectivity graph for the regions, in which the nodes denote regions, and we draw an edge between any two nodes if and only if there is a path from one region to the other that passes through no other region. We can search for a path in the graph from the node corresponding to the region that contains the start point to the node corresponding to (the region that contains) the goal point; the nodes on this path provide a list of regions (or streets), and we leave it to the robot (or some other plan refinement mechanism) to generate a detailed path. This clearly places more emphasis on the navigator, but otherwise is very similar to the Voronoi diagram technique.

As an example, Fig. 10.13(a) shows a set of regions for the example of Fig. 10.9, generated by forming a region boundary along a line of constant x whenever there is a suitable obstacle vertex, and Fig. 10.13(b) shows the corresponding connectivity graph. A suitable path can now be characterised by a graph path from the leftmost node to the rightmost node. This example illustrates the difficulty in producing 'sensible' regions with an automatic scheme, as the cost of moving between regions should not be regarded as the same in all cases (it clearly taking longer to move from the start to the neighbouring bottommost region than to the topmost region). This difficulty can either be resolved by ensuring that the regions are of (roughly) uniform size, or by building more competence into the graph search mechanism.

10.5 Recapitulation

At this stage it is useful to emphasis the breaking down of path planning into a number of steps:

1. *Problem Transformation.*
2. *Discretisation.*
3. *Path Search.*
4. *Refinement.*

Problem Transformation The first stage is to take the path-planning problem — of real shapes, moving under uncertain control, and with imperfect information — and t0 transform the problem into something that can be solved. For example: the moving-point transformation afforded by configuration space. This level of transformation preserves (most of) the metrical properties of the problem.

Discretisation There are too many possible paths to be considered, even after the simplifications in stage 1. A further transformation is required before the *"topology"* of the path can be considered — for example, by considering the visibility graph.

Path Search A path (or paths) is (are) generated using the data-structure defined during stage 2. A^* is commonly used, but other methods may be appropriate. The paths generated may, in fact, be impossible to execute with the robot, but they should capture the *topology* of the final path.

Refinement Path smoothing (or equivalent) may then be applied to produce a path that can be readily executed.

Breaking the path planning process down in this way makes it easier to understand the similarities and differences between different methods.

10.6 Rotations and Other Problems

So far we have limited our discussion to robots that are undergoing pure translations. The idea of C-space can be be applied to all robots, but many robots (and most vehicles) have more than two degrees-of-freedom (dof); they can change their orientation or z-coordinate as well. For any mechanical system we can write down a set of parameters that determines its configuration, such as the (x, y) coordinates for the translations we saw earlier, or the joint angles of a robot arm. As these (say n) numbers completely determine the position of the robot we can *theoretically* plot out, in n-dimensions, the configurations that cause a problem. This gives us the C-space obstacles, just as in the two dof case we saw earlier. However the mathematics of these spaces is quite complex, and in practise (and in particular for vehicles) the use of the types of C-space techniques that we have seen so far tends to be avoided. As an example, if we take the case of convex polygonal robot and obstacles, but now allow the robot to change orientation as well as translating, we have a 3 dof system, and the C-space obstacles are three-dimensional, with bounding surfaces made up of portions of the surface of a helix! Generating these C-space obstacles is considerably more involved than for polygons, and searching the subsequent free-space for paths is also very difficult. When planning paths for robot arms this work may be worthwhile, partly because arms tend to be reasonably accurate (and so it is possible to follow the paths generated), and partly because the clearances involved are tighter (making these methods necessary). For vehicle work, however, it is usually more practical to consider a class of solutions based on the concept of the potential field.

10.7 Following Your Nose

The sort of path planning algorithms discussed so far delight in trying to find 'optimal' solutions, as in the shortest path (for the visibility graph method), or the path that keeps furthest away from obstacles (for the Voronoi diagram method). For many vehicle problems concepts of optimality are much more vague: the ideal factory vehicle planner might be required to produce paths so that "the vehicle travels quickly, but without scaring the humans, and without needing to be recharged too frequently". It also should not take too long 'thinking' about where it should go. For such cases we might want to use a method called *generate-and-test* that works well much of the time, in which we have some (preferably simple) rules for generating a small number of potential paths; we then test the paths (to see whether a collision would occur), and rank them to find the best. The railway-track algorithm is the simplest case, in which the trains must follow the tracks, and collision detection is based on semaphores. The real problem comes when we have a greater choice of potential routes; the potential field paradigm is often used to resolve this choice.

For the simplest case of a potential field planner, let us return to the idea of C-space and the translation-only case. To find a path for the robot, it is sufficient to find a path for a point that lies within the free-space. Imagine that we place an isolated (say positive) electric charge at the starting position, that we charge up the C-space obstacles positively, and we fix a negative charge at the goal position. The isolated charge will then tend to be attracted to the goal position whilst being repelled away from the obstacles. Of course it is difficult to set up a physical model to achieve this effect, but it is quite simple to simulate the system. Each charge generates a potential field (so called because the field can be completely characterised by the value of the potential at every point), and the value of the total field can be computed numerically at any point in space. Thus the model is simulated by interactively computing the

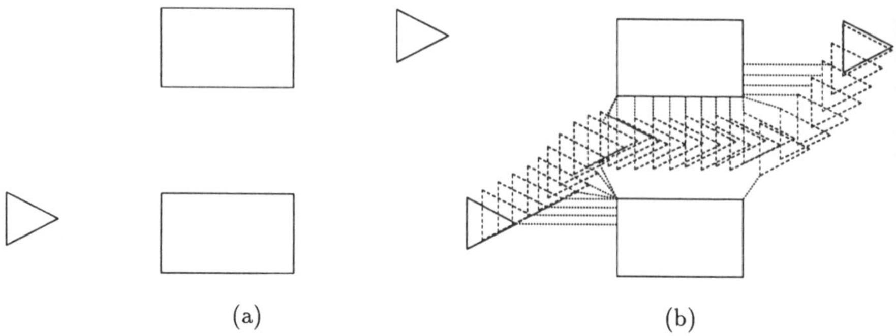

Figure 10.14: A simple potential field algorithm

force on the robot; moving the robot a little according to the force; and repeating until we reach the goal (or get stuck). For example, Fig. 10.14(a) shows a triangular robot that is being asked to move from left to right between a pair of positions in the presence of two rectangular obstacles, and Fig. 10.14(b) shows the snapshots generated by each iteration of the algorithm. The dotted lines with each snapshot join the closest points on an obstacle and the robot, and the force due to the potential field for each obstacle can be computed directly from knowledge of this vector, as it is lies along this line, with magnitude proportional to the inverse square of its length.

We can also plot out the value of the potential field, as in Fig. 10.15, which shows the field for the example of Fig. 10.14. (The net force on the robot at any point is given by the gradient of this surface.) The way that the potential field idea works can readily be appreciated by imagining the configuration of the robot as a ball-bearing, and letting it roll down the potential surface from the start position (top-left) to the goal (the dip at the bottom-right). In practice, we would want to also imagine the ball-bearing passing through some viscous medium to avoid it over-shooting the goal, but the biggest problem with the potential field method is in trying to avoid the 'ball' getting stuck at a *local minimum* in the field. In this case, a local minimum could easily be generated by making the repulsion due to the obstacles too strong, so that the 'ball' would get trapped between the two 'potential lumps' that correspond to the obstacles. Unfortunately we cannot just reduce the heights of these 'lumps' to avoid the problem, as then there is a risk of the repulsion not being strong enough to avoid a collision. Of course, in a particular case we could plot out the field and try to smooth out any problems by hand; but this does not make for a good automated path-planning system.

Despite this problem, progress has been made in recent years in providing useful path planning systems based around the idea of the potential field. Essentially, the

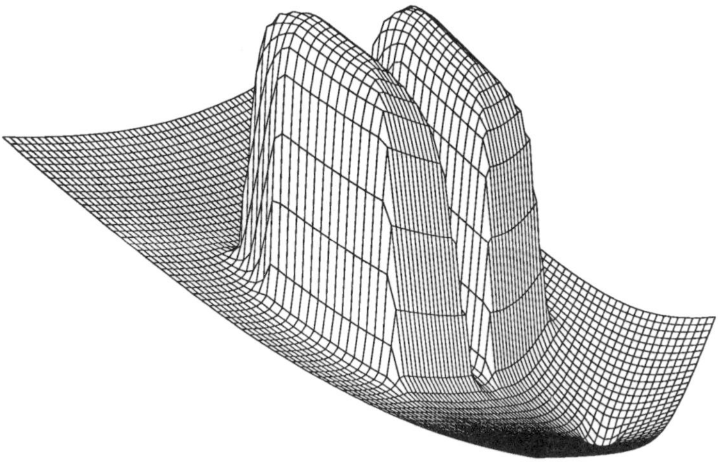

Figure 10.15: The potential field of Fig. 10.14

trick is to mix in some form of global sense to the field, either by sampling over a grid of points, or taking the time to design a good field, or by a careful analysis of a particular type of field. These topics will be discussed in more detail in Chapter 11, which describes a planner for robot vehicles that produces good paths for vehicles like the Oxford AGV.

Chapter 11

Path Planning for the AGV

Tony Hague

The basis of the potential field approach is as follows. Imagine that obstacles exert a repulsive force on the moving object, and goals exert an attractive force. The moving object will then be drawn towards the goal position, whilst simultaneously being forced away from obstacles. It can be shown that if the potential of regions within obstacles is set to infinity, collisions will never occur.

A well known problem from which this method suffers is the occurrence of local potential minima at locations other than the goal position into which the moving object may be drawn and become trapped. Various different methods of dealing with this problem have been devised, the more popular schemes being outlined in the discussion below.

The form of the artificial potential field may be defined in terms of the *minimum directed distance function* (Δ) [40] or the *minimum translational distance (MTD)* [46] between the moving object and the obstacles, in combination with a *potential function* of this distance. Thus in its simplest form, the potential field is well suited to reactive obstacle avoidance for mobile robots, on the basis of sonar range data, for example. Brooks [33] has employed such techniques within one layer of a subsumption architecture for control of a mobile robot; artificial forces which are functions of sonar range readings are combined by vector summation to give an obstacle avoiding force. Should this force have significant magnitude, avoiding manœuvres are performed.

Borenstein and Koren also use potential fields in combination with sonar data for real time obstacle avoidance for their mobile robot *CARMEL* [25]. However, their approach is to first combine the sonar data into a local uncertainty grid [60]. The repulsive forces are than made to be proportional in magnitude to the certainty value of occupied cells in the grid, and also to vary with the inverse-square of range.

The use of the uncertainty grid gives good immunity to the false readings which are common with sonar systems. Unfortunately, the discrete nature of the uncertainty grid causes substantial jumps in the magnitude and direction of the resolved forces, and so a digital low pass filter is needed to subsequently smooth the steer function. This filter, however, introduces a time lag betwen the sensing of obstacles and the avoiding action. In attempt to counter this a weighting is employed to increase the repulsion exerted by obstacles ahead of the robot. It therefore comes as no surprise that in their later work the authors report instability problems with the potential field method [26].

The potential field approach is used by Khatib [116] along with the transpose of the Jacobian to plan collision free paths for robot manipulators. Poon and Lawrence [164] employ a similar technique, except that their potential function rises abruptly to infinity at obstacle boundaries, rather than rising gradually in the manner of that employed by Khatib. This is necessary for storage requirements since the planning is done in a quantised configuration space, and also means that only a yes/no intersection test is required, rather than the calculation of a distance function such as Δ. It should be noted that the use of a potential function having an abrupt rise at obstacle boundaries will cause the moving object to scrape against obstacles. Local potential minima are escaped by filling the minimum with points of infinite potential. The path may subsequently be refined by repeated applications of the algorithm.

Krogh [121] introduces the concept of the *generalised* potential field. Here the field is a function of both the position *and velocity* of the moving object. This allows the potential experienced by the moving object to represent the time until a collision will occur, rather than simply the proximity of obstacles. This gives rise to better paths, since obstacles which have already been passed no longer influence the moving object, allowing it to head more directly to the goal location.

The control of a robot system using a potential field is commonly addressed by one of two approaches. The method employed by both Khatib and Krogh is a *force control* approach. The forces due to the field are treated as if they were real forces acting upon the robot, and the dynamics of the system are used to compute the motion such forces would produce. An alternative method which has the advantage of being computationally slightly simpler is to employ a gradient following technique, where the robot's motion is determined by the direction of steepest descent of the gradient of the potential field. The major disadvantage of this *hill-climbing* approach is that the dynamics of the system are not observed. A comparison of force control and gradient following approaches is given by Tilove [188]. This paper also considers the relative performance of the classical potential field (i.e., where the potential is a function of position only) and the generalised potential field method of Krogh described above.

The algorithms described above are suitable for path planning on the basis of *local* obstacle information. However, potential field like approaches have been also widely used to solve *global* path planning problems, where an *a priori* map of the environment is used to construct a potential function and a path to the goal location.

This brings the possibilty of constructing a potential field which is not simply due to local obstacles, but rather incorporates global knowledge to reduce or entirely avoid the problem of local minima. These algorithms are properly global planning algorithms. Those algorithms which are relevant to the current discussion of potential field based approaches are outlined below for completeness.

By topological considerations, Koditschek [119] argues that potential functions exist such that there is only one local minimum, the goal position. If such a function were employed with the standard relaxation paradigm, it would be possible to guarantee that the moving object could not encounter local minima. A class of such potential functions (called *navigation functions*) is presented for the restricted case of a point moving amidst circular obstacles within a disc shaped region. Later this work was generalised to address the case of "star worlds" [176].

Barraquand and Latombe [12] use a discrete grid representation of free space, and compute local minima free potential fields using a wave propagation approach. Such fields are used to guide the motion of robot manipulators having several degrees of freedom. Potential fields are constructed for control points placed at various locations on the manipulator. The influences of these fields are combined using the kinematics of the manipulator to give an overall potential. It is possible, however, for local minima still to occur in the combined potential. Escape from such minima is achieved using a probabilistically complete *Monte Carlo* technique. In order to address the problem of the large storage requirements and laborious search procedures inherent in techniques based upon the search of a discrete grid, a coarse grid is used initially. If the problem cannot be solved at this resolution, successively finer resolution grids are used until a solution is found, or a maximum resolution is reached, in which case the planner fails. This approach has the advantage that for relatively simple environments, it allows paths to be found very quickly.

The "Flexible Trajectory" approach described by Buckley [39] differs somewhat from the methods previously described. Instead of beginning with a start configuration and relaxing this through time, an initial hypothesised trajectory is deformed to minimise some cost function. In this case the cost function is based on Δ. Since Δ is defined within obstacles (where it has negative values), an initial trajectory which passes through an obstacle may be forced out. In the simplest case, a straight line connecting the start and goal locations will suffice as the initial trajectory. Whilst this method still may become stuck in local minima, it is less likely to do so than the usual relaxation paradigm. However, whilst the previously described methods ensure that collisions will not occur by virtue of the infinite potential within obstacles, this algorithm can give no such guarantee.

Hwang and Ahuja [102, 103] first derive a set of *minimum potential valleys* (MPV's) from a 2D environment by filling free space with maximal discs. The locus of the centres of these disks gives a graph which may be searched to find an initial trajectory, which is then subjected to a smoothing process. The path is then optimised using optimal control formulation to minimise a cost function based on path length and the integral of the potential experienced by the moving object. In addition to this optimisation, several heuristics are employed to allow suitable orientations of the moving object to be determined.

The preprocessing step used to compute the minimum potential valleys is quite costly but gives the benefit that this algorithm is immune to "getting stuck" in local potential minima. The method used for computing the MPV's is rather crude, and is particularly slow when filling long narrow corridors. This is somewhat unfortunate if the algorithm is to be used for planning paths within a warehouse, for example, where such corridors are likely to be a common feature.

It is interesting to note that the minimum potential valleys of Hwang and Ahuja are in fact a straight line approximation to the symmetric axis transform [21, 22], a structure which is closely related to the Voronoi diagram and is defined in terms of the locus of the centres of a maximal discs.

11.1 Potential Field Path Planning

In the following sections, the implementation of a basic potential field path planner is described. This algorithm may be used to find paths for a polygonal moving object amongst polygonal obstacles where the moving object has two translational degrees or freedom only.

An extension of this planner is then introduced, which allows paths to be found in the case where the moving object is also allowed to rotate, without resorting to the computationally expensive method of computing a three dimensional configuration space.

11.1.1 Defining the Form of the Field

For the sake of consistency, we will define here the form of the potential field which will be adopted for all subsequent examples. Let the potential due to an obstacle at any point in free space be proportional to the reciprocal of the minimum distance from that point to the obstacle. This gives rise to a resultant force upon the moving object which follows the Newtonian inverse-square law, acting directly away from the obstacle. Since only the resultant forces are needed for the algorithms presented here, the potential field itself need not be explicitly calculated. Note that the magnitude of the force is influenced by the range to the obstacle alone, and does not depend upon the dimensions of the obstacle. It is desirable that the clearance maintained from all obstacles be independent of their size. It is also necessary to prevent the field due to large obstacles (e.g, a corridor wall) effectively masking smaller obstacles (such as a box obstructing the corridor). The forces due to multiple obstacles will be combined arithmetically by vector summation. In order to guide the moving object towards the goal, an attractive force of constant magnitude acting in the goal direction is used. The combination of this force with the obstacle forces described above is used to guide

$$
\begin{aligned}
&\mathbf{x} \leftarrow \mathbf{x}_s \\
&\textbf{repeat} \\
&\quad \mathbf{F}_{repulse} \leftarrow (0\ 0) \\
&\quad \textbf{foreach } o \in O \textbf{ do} \\
&\quad\quad \mathbf{\Delta}_m \leftarrow \mathbf{\Delta}(M, o) \\
&\quad\quad \mathbf{F}_{repulse} \leftarrow \mathbf{F}_{repulse} + \widehat{\mathbf{\Delta}_m}/|\mathbf{\Delta}_m|^2 \\
&\quad \textbf{endfor} \\
&\quad \mathbf{F}_{attract} \leftarrow \widehat{\mathbf{x}_g - \mathbf{x}} \\
&\quad \mathbf{F}_{res} \leftarrow \mathbf{F}_{attract} + a\mathbf{F}_{repulse} \\
&\quad \mathbf{x} \leftarrow \mathbf{x} + s\widehat{\mathbf{F}_{res}} \\
&\textbf{until } (\mathbf{x} \approx \mathbf{x}_g) \textbf{ or } (|\mathbf{F}_{res}| \approx 0)
\end{aligned}
$$

Figure 11.1: Basic algorithm to follow a potential field

the moving object to the goal position whilst simultaneously avoiding obstacles.

11.1.2 Translation Only

We first consider a simple algorithm which attempts to find paths for the case of an object moving under translation only. Although the two dimensional case is considered here, the approach described may be readily generalised to the three dimensional case.

Consider a polygonal object M moving with two translational degrees of freedom in a workspace W in the presence of a finite set of polygonal obstacles O. Let the position of M be denoted by vector $\mathbf{x}$. Let the initial configuration be $\mathbf{x}_s$, and the goal configuration be $\mathbf{x}_g$. The following algorithm employs a potential field of the form described above to find a collision free path. For simplicity, a hill-climbing approach will be adopted for controling the motion of M, as shown in Fig. 11.1, where $\mathbf{\Delta}$ is a translational vector realising the minimum directed distance function [40, 39], and $\hat{\mathbf{v}}$ denotes the unit vector $\mathbf{v}/|\mathbf{v}|$. This minimum distance may be calculated quickly if configuration space obstacles are computed as a preprocessing step [46]. An alternative method would be to first transform the problem into that of navigating a point in configuration space, the difference between the two approaches being purely conceptual.

The constant a controls the extent to which obstacles repel M. A suitable value for a may be shown to be ϵ^2, where ϵ is the nominal clearance between M and obstacles. This follows from considering the point at which the goal attraction and obstacle repulsion forces are equal; closer to obstacles than this distance, the obstacle repulsion will dominate. The constant s defines the distance moved in each iteration of the algorithm. Clearly it is desirable that $s < \epsilon$ to avoid collision.

Since a potential function is used which is infinite within obstacles, collisions will not occur. However, the algorithm may fail to find a path due to the occurrence of local minima within the potential field. This is indicated by the termination of the loop with $\mathbf{x} \neq \mathbf{x}_g$. (Note that the tests for termination must include a degree of uncertainty, as indicated by the use of '$\approx$' in the tests, such as $|\mathbf{x} - \mathbf{x}_g| < \epsilon|$.) Various means could be employed to allow the moving object to escape from such minima, such as "filling" the minimum with a region of high potential [164] or by application of a Monte-Carlo approach [12].

11.1.3 With One Rotational Degree of Freedom

The algorithm described above can find paths for a polygonal moving object in an environment of polygonal obstacles (provided a local potential minimum is not encountered). If this technique is to be used to plan motions for a mobile robot, it must be extended to allow for a rotational degree of freedom.

The following describes a means of achieving this. The concept upon which the algorithm is based is somewhat intuitive, being arrived at as follows. It has been illustrated in the previous section that paths may be planned using a potential field approach, using a resultant force to direct translational motion. If this is so, *is it not possible for rotational motion to be determined by taking moments?* Consider an object moving along a minimum potential valley. If it is oriented such that the twisting moment due to the potential field is minimised, the effect will be to align the major axis of the moving object along the MPV. This is exactly the effect which is sought in [102] and [187] by the use of various heuristics.

In order to compute the moment acting upon the moving object due to the potential field, it is necessary to consider the effect of the field upon a polygon in operational space, rather than a point in configuration space. In order to simplify this computation, we first note that it is sufficient to consider the *boundary* of the moving object only, rather than the entire area. Initially, we simplify the computation still further by considering only a set of *control points*, uniformly distributed around the boundary. In §11.2, we will discuss a way of avoiding this by employing analytic solutions for the force and moment.

Let $\mathbf{c}$ denote the centroid of the moving object M, and P be a set of vectors describing the position of a number of control points uniformly distributed about the boundary of M relative to $\mathbf{c}$. The previous algorithm may be modified to include this extension, as shown in Fig. 11.2. The constant b controls the extent of the rotation. For the hill-climbing approach used here, this constant was determined empirically. It should be emphasised that although hill-climbing is adopted here, there is nothing to prevent the use of a force control approach, with the rotation being determined by consideration of the moment of inertia of the moving object. The moment of inertia might be approximated by considering the boundary of the moving object as a framework of thin rods, for which the moment of inertia about $\mathbf{c}$ may trivially be computed with the aid of the parallel axes theorem. This would avoid the need for empirically derived constants.

$\mathbf{x} \leftarrow \mathbf{x}_s$
$\theta \leftarrow \theta_s$
repeat
  $\mathbf{F}_{repulse} \leftarrow (0,0)$
  $moment \leftarrow 0$
  foreach $\mathbf{p} \in P$ **do**
    foreach $o \in O$ **do**
      $\mathbf{\Delta}_m \leftarrow \mathbf{\Delta}(\mathbf{c}+\mathbf{p}, o)$
      $\mathbf{F}_p \leftarrow \widehat{\mathbf{\Delta}_m}/|\mathbf{\Delta}_m|^2$
      $\mathbf{F}_{repulse} \leftarrow \mathbf{F}_{repulse} + \mathbf{F}_p$
      $moment \leftarrow moment + (\mathbf{p} \times \mathbf{F}_p).\mathbf{k}$
    endfor
  endfor
  $\mathbf{F}_{attract} \leftarrow \widehat{\mathbf{x}_g - \mathbf{x}}$
  $\mathbf{F}_{res} \leftarrow \mathbf{F}_{attract} + a\mathbf{F}_{repulse}$
  $\mathbf{x} \leftarrow \mathbf{x} + s\widehat{\mathbf{F}_{res}}$
  $\theta \leftarrow \theta + b * moment$
until $(\mathbf{x} \approx \mathbf{x}_g)$ **or** $(|\mathbf{F}_{res}| \approx 0)$

Figure 11.2: Using moments to allow rotations

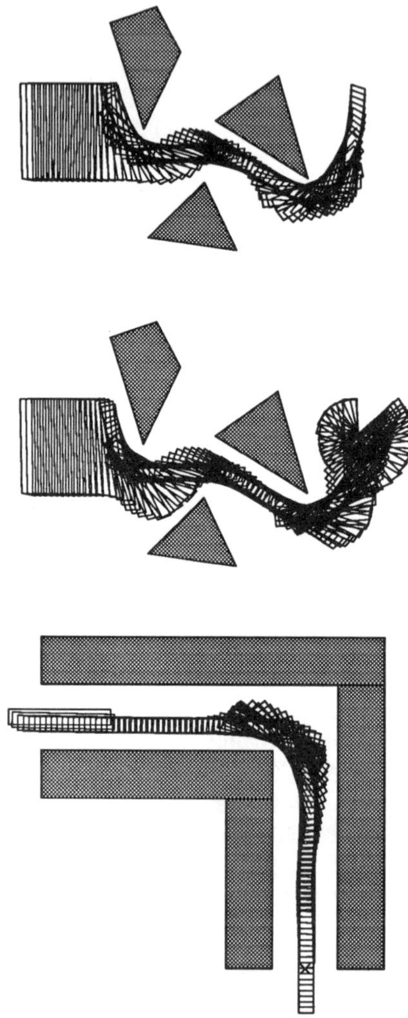

(a)

(b)

(c)

Figure 11.3: Example paths planned for an object moving with two translational and one rotational degree of freedom

Example paths generated by the above algorithm are shown in Fig. 11.3. In all the examples the start location is at the left. Thirty control points were used in each case. Fig. 11.3a shows a solution to the problem which could not be solved by the translation-only algorithm of §11.1.2. Fig. 11.3b was generated using a variant of the algorithm which allows the goal orientation to be specified. An extra component is added to the moment computed which draws M into the required goal orientation, in a manner analogous to the use of an attractive force in the goal direction to guide translational motion in the direction of the goal position. Finally Fig. 11.3c shows the problem of moving a rectangular rod around a 90° corner. Note that the motion along a straight corridor is smooth and in no way oscillatory, with only a slight overshoot in orientation in coming out of the corner.

11.2 Calculation of the Influence of An Artificial Potential Field upon a Polygonal Moving Object

It is possible to replace the control points used in the previous section by integrating over the boundary of the robot instead. Let the force acting on an infinitesimal length δl of the boundary of the moving object M at point $\mathbf{P}$ in space due to the artificial potential field of obstacle $o \in O$ be defined as follows :-

$$\delta \mathbf{F} = \frac{\rho}{|\mathbf{\Delta}|^2}.\widehat{\mathbf{\Delta}}$$

where $\mathbf{\Delta}$ is the vector realising the minimum distance between point P and obstacle o, and ρ is a constant mass / unit length of the boundary of M.

Recall that the field was chosen to be such that the potential at any point due to all obstacles in the environment is the sum of the potentials due to each obstacle, hence the total force and moment due to obstacle repulsion is simply the sum of the forces and moments respectively due to each obstacle $o \in O$. Furthermore, since the force acting is dependant upon the minimum distance vector $\mathbf{\Delta}$, it is possible to divide the free space around each obstacle o into a set of regions, such that within each region, the potential at each point is dependant only upon one vertex (i.e, R_2, R_4, R_6, R_8 and R_{10}) or edge (i.e, R_1, R_3, R_5, R_7 and R_9) of o. These regions are illustrated in Fig. 11.4. The boundaries of the regions are lines perpendicular to each obstacle edge, and passing through obstacle vertices. Regions associated with an obstacle edge are also bounded by the edge itself. By clipping the boundaries of M against these region boundaries, it is possible to divide the boundaries of M into a set of shorter line segments, each of which lies wholly within one of these regions. Therefore there are two cases for which a solution is required. We must determine the force and moment

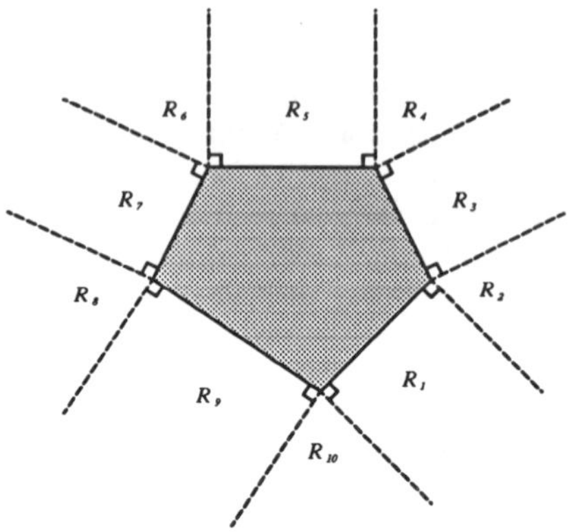

Figure 11.4: Dividing the space around obstacles into zones to allow the force and moment due to the potential field to be calculated using a closed form solution. Within each of the zones shown in dotted lines, the repulsive force is affected only by the corresponding obstacle edge or vertex

due to an obstacle edge, and due to an obstacle vertex. These are derived in [82]; here we only present the results.

Case 1 : Computing the resultant force and moment upon the moving object due to an obstacle edge

Refering to Fig. 11.5:

$$\mathbf{F} = \frac{\rho\,|\mathbf{u}|\widehat{\mathbf{r}_1}}{|\mathbf{r}_1||\mathbf{r}_2|} \tag{11.1}$$

where $\mathbf{u} = \mathbf{U}_2 - \mathbf{U}_1$. The moment m about an arbitrary point $\mathbf{C}$ is given by:

$$\begin{aligned} m &= (\mathbf{U}_1 - \mathbf{C}) \cdot \hat{\mathbf{p}}\,|\mathbf{F}| + \\ &\quad \mathbf{u}.\hat{\mathbf{p}}\,\rho\,|\mathbf{u}|\frac{1}{(|\mathbf{r}_2| - |\mathbf{r}_1|)^2}\left[\ln\left(\frac{|\mathbf{r}_2|}{|\mathbf{r}_1|}\right) - \left(\frac{|\mathbf{r}_2| - |\mathbf{r}_1|}{|\mathbf{r}_2|}\right)\right] \end{aligned} \tag{11.2}$$

Note that this equation cannot be evaluated when $|\mathbf{r}_1| = |\mathbf{r}_2|$, but that it is a straightforward to switch over to evaluating Taylor's expansion near this point instead.

Case 2 : Computing the resultant force and moment upon the moving object due to an obstacle vertex.

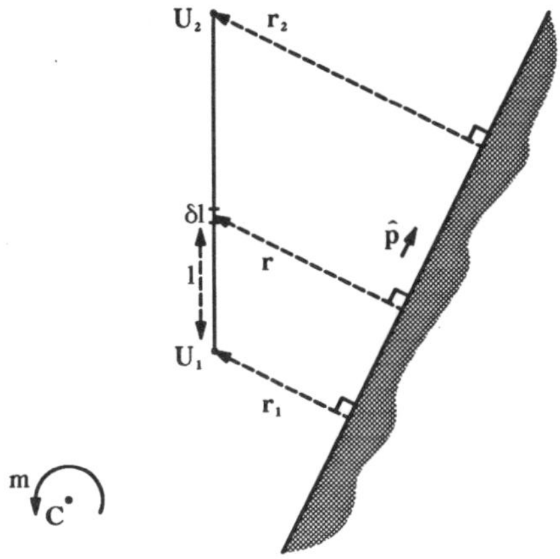

Figure 11.5: Calculating the influence of an obstacle boundary upon an edge of the moving object

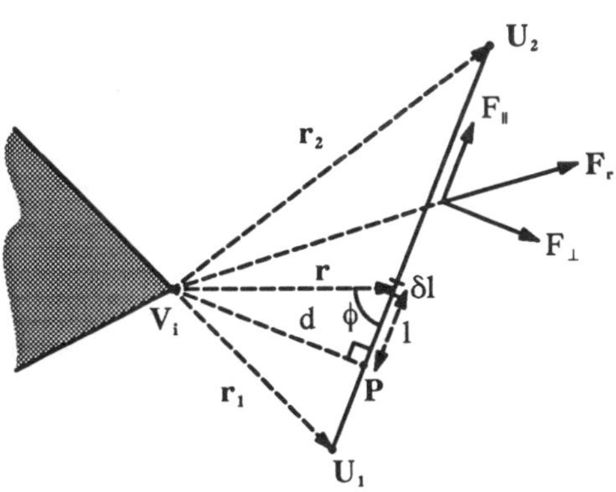

Figure 11.6: Calculating the influence of an obstacle vertex upon an edge of the moving object

From Fig. 11.6 it is clear that the line of action of the resultant force must pass through the obstacle vertex $\mathbf{V}_i$. Therefore if the resultant force $\mathbf{F}_r$ can be calculated, the moment about an arbitrary point follows trivially. In this case, we consider two orthogonal components of $\mathbf{F}_r$, $F_{\|}$, the component parallel to the edge of M, and $F_{\perp}$, the perpendicular component.

$$F_{\|} = \rho\left[\frac{1}{|\mathbf{r}_1|} - \frac{1}{|\mathbf{r}_2|}\right] \tag{11.3}$$

$$F_{\perp} = \frac{\rho}{d}\hat{\mathbf{u}} \cdot (\widehat{\mathbf{r}_2} - \widehat{\mathbf{r}_1}) \tag{11.4}$$

From equations (11.3) and (11.4) both the force and moment can easily be computed.

Case 3 : Goal attraction

Finally it is necessary to compute the effects of goal attraction. The goal will be a point attractor, exerting an attractive force of constant magnitude ρa per unit length of boundary of the moving object, where a is a scalar constant used to control the relative weighting of goal attraction and obstacle repulsion. Two orthogonal components of the force will again be computed. The computation is similar to case 2.

$$F_{\|} = \rho a(|\mathbf{r}_2| - |\mathbf{r}_1|) \tag{11.5}$$

$$F_{\perp} = \rho a d\left[\sinh^{-1}(L_2/d) - \sinh^{-1}(L_1/d)\right] \tag{11.6}$$

Thus it is possible to solve the problem of computing the influence of the potential field upon the boundary of a polygonal obstacle by dividing the boundary into a set of line segments such that for each segment there is a standard result which can be applied to give the force and moment about an arbitrary point. These can be summed for all segments, and all obstacles, to give the total force and moment due to obstacle repulsion. The overall influence of the field is completed with the addition of the force and moment due to goal attraction computed from (11.5) and (11.6).

11.3 Potential Field Path Planning for Robot Vehicles

We address the problem of finding paths for a wheeled robot vehicle. In particular, the case of the car-like robot will be considered, that is, either a tricycle having two

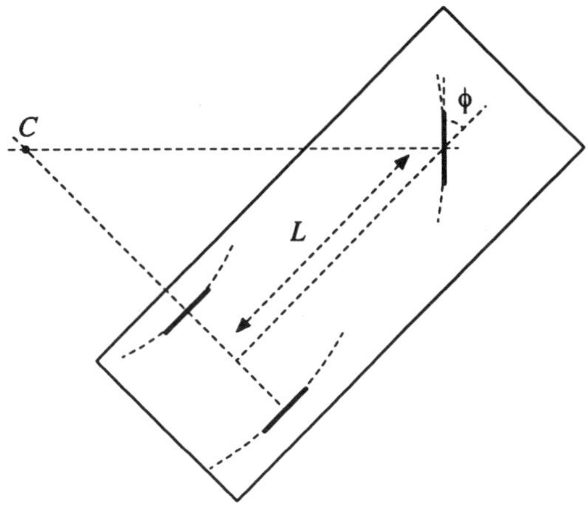

Figure 11.7: A three wheeled vehicle turns: at any instant, the three wheels lie tangent to circles centered at some point C, the instantaneous centre of gyration

rear wheels and a single steered front wheel, or a four wheeled vehicle where the front steered wheels are connected by a mechanical linkage such that its motion closely approximates that of a tricycle. Such a vehicle may not translate and rotate freely in the plane. Consider Fig. 11.7. The midpoint of the rear axle of the vehicle may only follow a path where the mid line of the vehicle lies tangent to the path at all points. The equations governing the motion of the vehicle are as follows:

$$\dot{x} = v_r \cos\theta \tag{11.7}$$
$$\dot{y} = v_r \sin\theta \tag{11.8}$$
$$\dot{\theta} = \frac{v_r}{L}\tan\phi \tag{11.9}$$

Here $[\,x\ y\ \theta\,]$ is the location of the mid point of the rear axle, and v_r is the velocity of that point on the vehicle. Alternatively, rather than using the centre of the rear axle, the equations might be written with the front wheel of the vehicle as the reference point thus:

$$\dot{x} = v_f \cos(\theta+\phi) \tag{11.10}$$
$$\dot{y} = v_f \sin(\theta+\phi) \tag{11.11}$$
$$\dot{\theta} = \frac{v_f}{L}\sin\phi \tag{11.12}$$

Here v_f is the velocity of the front wheel. This alternative formulation will be of use in later examples based upon the AGV, where the front wheel is both steered and driven. In the discussion here, however, we will use the first form given. From the equations (11.7) and (11.8) the kinematic constraint may be written thus:

$$-\dot{x}\sin\theta + \dot{y}\cos\theta = 0$$

Or alternatively:

$$[\,\dot{x}\ \dot{y}\,] \cdot [\,-\sin\theta\ \ \cos\theta\,] = 0$$

Since $[\,-\sin\theta\ \ \cos\theta\,]$ is a unit vector perpendicular to the mid-line of the vehicle, this simply states that the component of the vehicle's velocity in that direction is zero, i.e, the vehicle is not capable of making sideways motion. It is not possible to integrate this velocity constraint where both the steering angle ϕ and drive velocity v_r are arbitrary functions of time [130, 128]. Indeed, this velocity constraint does not restrict the $[\,x\ y\ \theta\,]$ configurations which are reachable by the vehicle (a proof of this is given in [130]). Such a constraint, which does not reduce the dimension of the configuration space but reduces the number of degrees of freedom of the moving object or mechanism, is said to be *non-holonomic.*

The existence of this non-holonomic constraint upon the vehicle's motion is problematic when planning paths which must respect the vehicle kinematics. Since the velocity constraint is not integrable, it cannot be transformed into a simple constraint on the geometry of feasible paths. Given steering angle $\phi(t)$ and drive velocity $v_r(t)$, it is not generally possible to find analytically $x(t)$, $y(t)$ and $\theta(t)$, or vice versa.

One possible solution is to find paths which are composed of arcs having a particular form, for which the equations of motion may be integrated. A popular method is to find paths which are composed entirely of straight lines and circular arcs. The effect of this is that for each path segment, $\tan\phi$ in (11.9) is constant. In the case where $v_r(t)$ is either constant, or varying in proportion to t (ie, the velocity is ramping up or ramping down) the equations can be integrated.

Wilfong [196] finds paths composed of straight line segments along precomputed lanes connected by turns having the form of circular arcs. Suitable radii for the arcs are selected by considering the *critical radii*, at which the vehicle would come into contact with an obstacle feature, and hence computing the intervals of the radius for which collisions will not occur. Manual intervention or heuristic methods are necessary to select the particular radius within these free intervals which will be used.

An alternative approach is used by Fraichard [73], where the radius of curvature of the corners is kept constant, and the location of the centre of the circle is chosen to give a collision free path. This is done by computing the *curvature centres space*, being the set of locations for the centre of the circular arc such that the turn will be collision free. Successive turns are connected by bi-tangent straight line segments.

In order to find minimum length paths, Jacobs and Canny [107, 108] first show that such a path may be assembled from canonical components composed of straight lines and circular arcs. Since each canonical segment may be described by a small number of parameters, it is possible to construct the configuration space for these path segments, being of dimension equal to the number of parameters of the path segment. Thus a sequence of collision free path segments can be assembled. Although of theoretical interest the practicality of this method is somewhat limited, since minimum length paths necessarily involve contact with obstacles, and a point vehicle moving amongst polygonal obstacles is assumed.

Barraquand and Latombe [11] use a search of configuration space to find collision free paths. For a given $[\,x\ y\ \theta\,]$ configuration, six successors are generated by considering a small motion with the velocity v set to one of $\{-v_0, v_0\}$ and steer angle ϕ set to one of $\{-\phi_{max}, 0, \phi_{max}\}$. By thus setting the velocity and steer angle constant, the incremental change in $[\,x\ y\ \theta\,]$ may be easily computed. The search is guided by a cost function to minimise the number of reversals of the direction of travel, and the total path length. In order to ensure a collision free path, intersections between the robot and obstacles are tested for at each configuration attained. Using this approach, paths are found for both a simple robot vehicle, and a vehicle towing a trailer.

A limitation common to all the algorithms mentioned above is that although the paths found are curves of class C^1 (that is, continuous in the first differential[1]) the second differential, and hence the curvature of the path, has discontinuities at the transitions between motion along straight line and circular arc segments. The physical significance of this is that the demanded steering angle along the path is not continuous. Since it is not possible for a real vehicle to steer infinitely quickly, the vehicle is forced to stop to steer, if such a path is to be followed precisely.

A possible solution to this, considered by Pin and Vasseur [163], might be to use transitional trajectories to connect straight line and circular arc segments together such that the resulting path is of class C^2. If the turning radii used are restricted to a discrete set, these transitional trajectories might be precomputed off line. A more elegant solution, perhaps, would be to compute paths which rather than using circular arcs as a primitive component, instead employ *clothoids*. A clothoid is a curve for which the curvature function $c(s)$ is linear with respect to arc length s. That is:

$$c(s) = ks + c_0$$

Thus for a path composed of straight segments and clothoid curves, the demanded steer angle is *piecewise linear* with respect to path length.

Kanayama and Miyake [113] study the use of clothoids to specify smooth trajectories to be followed by the control system of a robot vehicle. Given a set of via points or *postures* of the form $[\,x\ y\ \theta\,]$, their algorithm computes a smooth trajectory composed of straight lines and clothoid curves which connects the given postures. Where more than one trajectory might be found, rules are applied to ensure a unique solution.

[1]A curve is said to be of class C^n if it is differentiable n times and if the nth derivative is continuous.

Laumond has proved [130] the existence of a collision free trajectory for a car like robot subject to non-holonomic and steering angle constraints between two configurations if they are included in the same connected component of free configuration space. This result has lead to the development of various planners. In [129] a path is first found by searching a cellular decomposition of free space. A path through these cells, composed of straight lines and turns on the spot, is then computed. Finally the path generated is smoothed using clothoids. In general, such an algorithm should be expected to produce paths with many reversing manœuvres. In [109], an initial path which does not respect the kinematic constraints is computed by any of several techniques. In order to derive a path for the non-holonomic vehicle from this, the results of Reeds and Shepp [168] for the shortest path between two configurations is used. The path is recursively subdivided until the shortest paths connecting the intermediate points are collision free. This gives a path which is collision free, and respects the kinematic constraints, but again has many manœuvres. This path is then "optimised", by iteratively selecting pairs of randomly chosen intermediate points along the path, and constructing the shortest path between the two configurations. If the shortest path is collision free, it is substituted for the original segments between the two configurations.

An unusual approach is adopted by Tournassoud and Jehl [190, 189]. A decomposition of free space into generalised cones similar to that of Brooks [32] is used. The problem of turning corners is addressed by showing that if the trajectory generated by sliding continuously along the outer wall is not collision free, then no smooth collision free trajectory exists. The planner described finds cornering trajectories determined by the sliding constraint, and the vehicle's kinematic constraint. However, the case where the steering angle is constrained is not addressed.

11.4 Local Path Planning for Robot Vehicles Using a Potential Field Approach

The methods described above adopt a global approach to path planning. Here we consider a local potential field based approach to vehicle path planning, to fit within the two level framework we have adopted. The method described here is based upon the results for the influence of an artificial potential field upon a rigid body. The use of these results allows the geometry of the vehicle to be taken into account. This is somewhat rare amongst existing potential field planners, which often approximate the vehicle as a point. Alternatively, where obstacle data is gathered solely from a ring of sonar sensors distributed around the edge of the vehicle, advantage may be taken of the implicit configuration space transformation resulting from this sensor arrangement (i.e., that the range readings returned by the sonars are equivalent to the range information that would be measured in the configuration space of the vehicle). The method adopted here allows paths to be found on the basis of a geometric model of the environment, which may represent à priori knowledge of the workspace, or may be the result of processing sensor data, or a fused combination of these. The consequence of using the results in this way is that paths may be found through significantly tighter environments than would normally be possible using a potential field approach.

As the vehicle is not free to translate and rotate freely, the forces and moments due to the action of an artificial potential field may not be applied to the vehicle directly to cause motion, as was the case for an unconstrained polygonal moving object in the examples. In much of the existing literature on the subject of motion planning for mobile robots using potential fields, the issue of how the forces computed might be used to control the motion of a non-holonomic vehicle is not addressed. The problem may be evaded entirely by adopting a vehicle which is not subject to non-holonomic constraint (e.g, a vehicle which maintains a constant orientation, and has direction of travel and speed as its control parameters). The Cybermation K2A platform [106] (which forms the basis of the University of Michigan's *CARMEL* mobile robot employed by Borenstein and Koren) is of this type, as was the vehicle employed by Connell at MIT [53]. The problem may alternatively be simplified by using a "turtle" type vehicle which may turn on the spot, such as the *YAMABICO* family of mobile robots developed at the University of Tsukuba [198].

11.4.1 Guiding a Car-like Robot Using Potential Fields

One possible approach to the problem of controlling the motion of a car-like robot would be to resolve the resultant force into two components, one in the direction the vehicle is heading, and one perpendicular to it. The former is then used to control the speed and direction of travel of the vehicle, the latter the steer angle. However, consider Fig. 11.8, which once more shows a vehicle turning a corner about point C. The dashed lines in this figure show the paths taken by the corners of the vehicle. It may easily be seen that turning causes points lying to the rear of the vehicle's rear wheels to swing *outward*. Thus, if the scheme suggested above were followed, the effect of the repulsive field due to an obstacle towards the rear of the vehicle would be to bring the vehicle *closer* to that obstacle. This causes poor behavior when turning corners in close proximity to obstacles. An alternative scheme that we have developed is to use the *moment* of the forces due to the field about the reference point R, sited at the midpoint of the vehicle's rear axle, to control the steering angle of the vehicle. The point R is chosen since it lies at the intersection of the line passing through the axis of the rear wheels and consequently the centre of gyration C, and the axis of symmetry of the vehicle. This method gives a steer function which is in the correct sense for both obstacles in front of, and behind, the rear axle.

For the examples shown here, the steering angle is made to vary in proportion to the moment experienced by the vehicle. (The automatic determination of a suitable constant multiplier is discussed in [82].) This is used within an iterative algorithm, which computes the moment upon the vehicle and adjusts the steer angle at set dis-

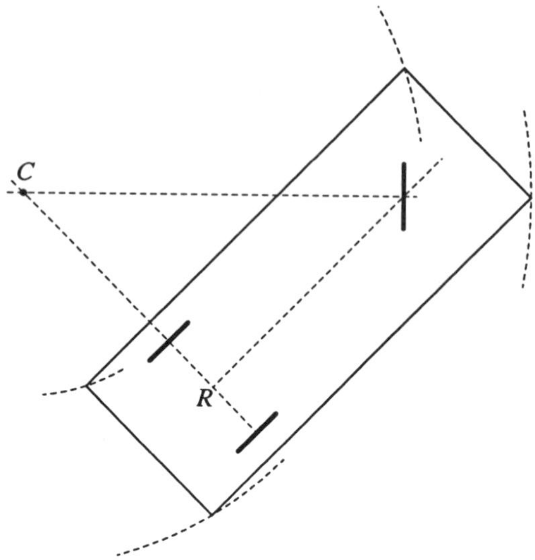

Figure 11.8: A three wheeled vehicle turns about point C. The dashed arcs show the motion of the corners of the vehicle

tance intervals. The work to date has focussed on off-line planning, and thus the motion of the vehicle is predicted from a model of the vehicle kinematics, approximating the drive velocity and steer angle as constant over each increment. Integrating (11.9), and assuming v_r and ϕ to be constant over time period τ, gives:

$$\theta_n = \theta_{n-1} + \frac{v_r \tau \tan\phi}{L}$$

Setting $v_r \tau = \delta$, where δ is a small distance along the path, and integrating (11.7) and (11.8) we have:

$$\theta_n = \theta_{n-1} + \frac{\delta \tan\phi}{L} \qquad (11.13)$$

$$x_n = x_{n-1} + \frac{L}{\tan\phi}(\sin\theta_n - \sin\theta_{n-1}) \qquad (11.14)$$

$$y_n = y_{n-1} - \frac{L}{\tan\phi}(\cos\theta_n - \cos\theta_{n-1}) \qquad (11.15)$$

If the algorithm were used for real-time reactive planning, this prediction of vehicle position would be combined with sensor data to provide an estimate of the vehicle's position.

In summary, the basic algorithm proceeds as follows:

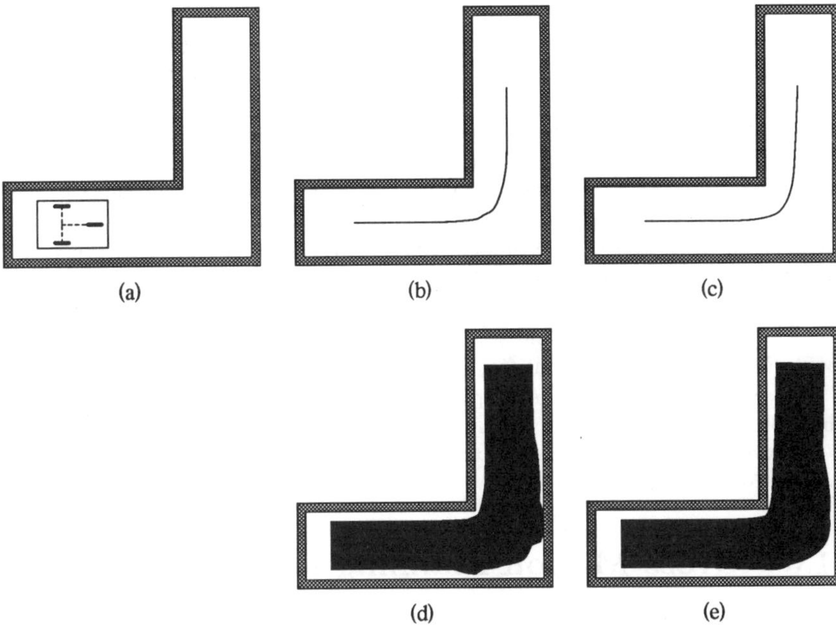

Figure 11.9: Turning a corner using potential field methods; (a): the vehicle is at the start location; (b): a path of reference point R found by using the resultant force to control the steer angle; (c): the result of using the moment of the forces upon the vehicle to control the steer angle. (d) and (e) show the area swept out by the vehicle following the paths in (b) and (c) respectively

repeat
    compute moment m using eq. (11.1) $\cdots$ (11.6)
    set $\phi = k * m$
    move small increment δ,
    update vehicle position $\mathbf{x}$ using eq. (11.13) $\cdots$
(11.15)
until $\mathbf{x} \approx \mathbf{x_g}$

Here $\mathbf{x}_g$ denotes the goal location, as before.

Fig. 11.9 shows a three wheeled vehicle turning a corner. Fig. 11.9(a) shows the vehicle at the start location, with the position of the wheels illustrated. For these examples, the drive velocity v_r is kept constant. The vehicle moves under the influence of obstacle repulsion alone; no attractive goal location is specified. The vehicle steer angle is determined from the potential field. The two different approaches to guiding the vehicle discussed earlier are shown here for comparison. Fig. 11.9(b) and Fig. 11.9(d) show the path generated by using the component of the resultant force in the direction perpendicular to that in which the vehicle is facing to control the steer angle. In Fig. 11.9(c) and Fig. 11.9(e), the moment about the mid-point R of the rear axle is used. Both the path taken by the reference point R, and the area swept out by the vehicle, are shown for both cases.

Notice that in (b) the path has a small peturbation as the vehicle first turns too sharply into the corner, and has to make subsequent correction. The effect is readily seen in the area swept out by the vehicle shown in (d); the repulsive forces due to the bottom wall cause the vehicle to steer (incorrectly) more sharply left, bringing the rear of the vehicle nearer to the wall. Later, the repulsion of the inside wall causes the vehicle to steer to the *right*, causing the near collision with the right wall seen in (d). The path shown in (c) and (e), found using the moment of the obstacle forces about R to guide the vehicle, is free from these flaws.

It is worth note that this example represents a difficult problem. An algorithm which plans cornering manœuvres by finding a circular arc connecting the mid lines of the two corridors which meet at that corner would fail here, since there exists no collision free path of that form.

11.5 Gross Motion Planning for Mobile Robots

It is not necessary for the paths generated by the global planning stage to take into account detail of the geometry of the vehicle, or the kinematic constraints to which it is subject. These issues will be addressed by the "moments" algorithm. Neither is it necessary for the global planner to produce very detailed paths; a simple path composed of straight line segments is sufficient. However, it is most desirable that the space over which the search for a path is performed be minimised, so that paths may be found more quickly.

A method based upon the Voronoi diagram could be used, and indeed such an approach may be most appropriate in highly unstructured environments. However, for the more structured indoor environment, the paths generated from the Voronoi diagram may often be a poor starting point. Consider Fig. 11.10. Here a straight path between the start and goal locations, as shown on the left, represents a good initial path. The path derived from the Voronoi diagram is shown on the right. This path is made up of several nodes and connecting arcs, some of which are parabolae. The overall path is discontinuous in the first derivative with respect to path length. The nodes do not make suitable via points for a potential field path following module, as they draw the robot into "side roads".

The first of our algorithms is intended to address this case. A routemap approach is employed. For AGV systems currently in use in industry which employ routemaps the maps are manually derived. The algorithm presented here is largely intended to automate that process, taking as input a 2D plan of the environment. This plan could

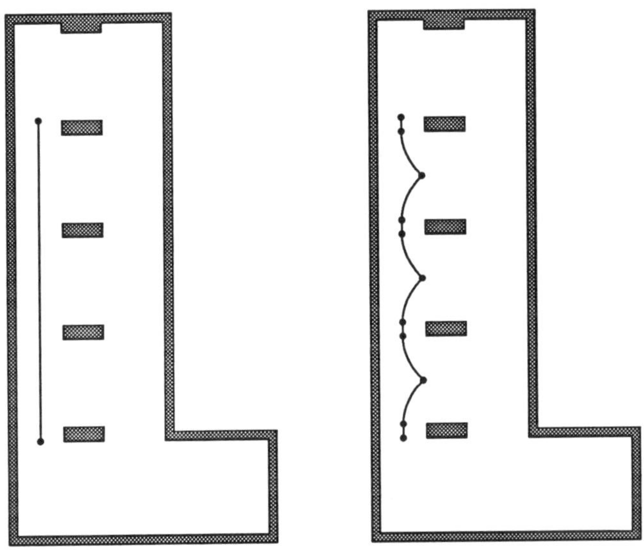

Figure 11.10: Example paths. Left: a straight line path connecting two points in free space. Right: the path which is yielded by searching the Voronoi diagram with the same two terminal points

be either manually entered or, as is the case for the environments used in the various examples shown here, derived by projection from a 3D solid model.

Much of the labour in computing a routemap is in determining the adjacency relationships between obstacles, so that path segments passing between them may be constructed. In [123], for example, this is done by forming the convex hull of pairings of obstacles, then testing for intersections of this convex hull against all other obstacles. In the event that no intersection is found, the region between the pair of obstacles is used as the basis of a *freeway*. Clearly the time complexity of the naïve implementation of this algorithm will be $O(n^3)$, where n is the total number of obstacles in the environment. The first algorithm presented here, like that of Singh and Wagh [184], exploits the fact that man-made environments such as factories and warehouses in which AGVs are employed often have, or may be closely approximated as having, a *Manhattan* geometry. That is, the environment may be represented by a model in which all obstacle boundaries lie in one of two mutually orthogonal directions. This is a generalisation of the environment of iso-orientated rectangles, since obstacles which are not a simple rectangle may be used without need for decomposition into rectangular components. In this case adjacency information may be quickly extracted by *sorting* the obstacle boundaries. Our method differs from that of Singh and Wagh in that we will not decompose free space into a set of maximal overlapping rectangles, but will instead identify rectangular regions of maximal *length* for use as freeways or *corridors*; the method is described in Chapter 13.

11.6 Summary

We have shown how it is possible to use the moment of a potential field to guide a robot without explicitly computing its configuration space. Further by careful consideration of the way in which the AGV moves it is possible to efficiently guide it, despite its car-like steering. We have also demonstrated the use of global information (in the form of a route-map) to provide an alternative way of providing a 'moving goal' for the vehicle, resulting in a path planner that is fast enough to be used for real applications.

Chapter 12

Task Planning

Jian Peng and Stephen Cameron

The work reported in this chapter is motivated by the fact that there is a gap between 'traditional' work in planning (as championed by the AI community) and robot planning in the real world: the former has concentrated on good knowledge representation techniques but in idealised domains, while the latter has to deal with the rich geometric knowledge arising from the environment. To bridge the gap, we use a hierarchical approach to the construction of robot spatial reasoning systems, with different levels in the hierarchy adopting different representations to handle different types of planning knowledge. A prototype of such a system has been implemented in the POPLOG environment where programs coded in different languages can be easily mixed together.

The problem domain involves the AGV moving pallets [31]. That is, we wish to introduce the AGV into a space into which a number of loaded pallets have been positioned in an irregular manner. The AGV will have a fork-lift attachment (simulated in our case!), and has to identify the pallets (as in Chapter 9), compute their orientations, and plan how to acquire the pallets using the fork-lift. Figure 12.1 shows such a working environment using the output of ROBMOD—the solid modeller employed to implement part of the world model Chapter 13. The loaded objects and obstacles are not shown.

The palletiser [162] is unlike many path-planning systems in that it has to plan paths for the AGV that bring it into (gentle) collisions with objects (*i.e.* so that the forks get inserted into the pallets). In the general case path planning involving contacts is a hard problem; what makes it easier in our case is the small class of objects with which we wish to contact. Thus our plan of attack is roughly schema-based; we use common-sense (rule-based) reasoning to suggest outline plans, which

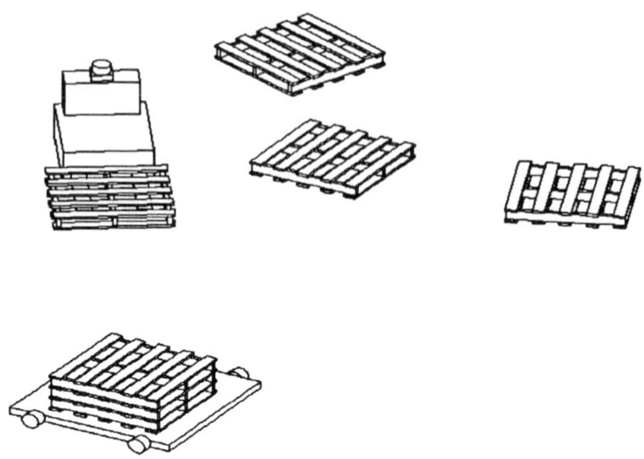

Figure 12.1: The AGV at work

are then examined using machinery that has a more thorough knowledge of matters geometrical. Through this work we wish to identify geometric functions that could form the interface between the two styles of reasoning, and that could be used to provide an even more versatile geometric reasoning facility.

12.1 Overview

Figure 12.2 shows the architecture of the palletiser subsystem. The world model contains both a fixed *a priori* map of the environment, plus modifications suggested by the various sensors of the AGV (such as described in Chapter 9). The palletiser basically schedules the AGV to collect (multiple) pallets. It also incorporates the extra constraints imposed by pallet acquisition and, together with the path planners, brings the AGV into contact with pallets. Note that in this design there is a pair of simple-minded path-planners within the palletiser; these planners have a relatively low-level of competence but, as a result, more generality than the external, 'expert' path-planner, which may need to be changed for different tasks and vehicles. The following sections describe the system in a top-down manner, thus: the specification of the basic scheduler; the crudest (and fastest) path planner; various refinements to the scheduling process; the better internal path planner (based on the generalised Voronoi diagram); and the final system.

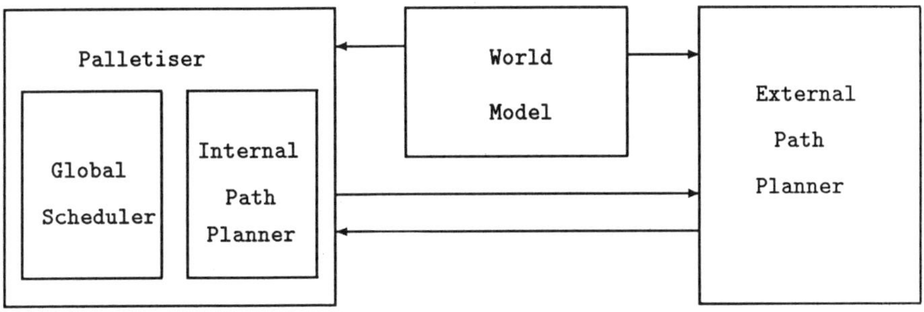

Figure 12.2: The architecture of the palletiser subsystem

12.2 The Scheduler

The AGV is required to collect multiple pallets, and so the scheduler constructs the order in which the AGV accesses the goals. To complete the specification of the problem we imagine that the AGV starts at a standard position (the 'processing centre'), collects a pallet, and returns with it to the processing centre, at which point the collected pallet is removed from further consideration. (This particular choice of the start and goal positions is only for our convenience when setting up the problems; the resultant planner can easily handle more general cases.)

The problem can be regarded as a repeating collision-free path planning process, with each iteration planning a path to collect one pallet. Each path consists of two segments: one for the AGV to approach the pallet from the processing centre, and the other for the AGV, loaded with a pallet, to go back to the processing centre. In most cases, if the combined shape of the AGV and a pallet can find a path to the processing centre, then the AGV can take the same path to the pallet, and so we only have to consider the returning segment of each path. In the following context, we use "the pallet has a path to the goal" to mean that "the combined shape of the AGV and the pallet has a path to the goal".

Note that the pallets themselves also serve as obstacles during the path planning stage. However, they are removable, and this results in a *constantly changing* geometric environment in which we are planning paths.

The PROLOG language [51] is an integral part of the POPLOG system, and so we used it to define the higher-level constraints. The basic problem can be specified in PROLOG thus:

```
findorder(Original,Result):-
        permutation(Original,Result),
```

```
        sortbypath(Result).
sortbypath([]).
sortbypath([P1|Prest]):-
        path(P1,Prest),
        sortbypath(Prest).
```

where permutation(X, Y) is true if X is a permutation of Y, and path(A, B) is true if A has a path to the goal without colliding with any element of B.

Calling the predicate `findorder(Original,Result)` therefore results in trying to find a permutation `Result` of `Original`, which satisfies `sortbypath`. `sortbypath` takes a list of pallets and evaluates to true if any pallet in the list is not blocked by those pallets after it (*i.e.* it has a path to the goal without colliding those pallets).

12.2.1 Improving the Solution by Geometric Reasoning

The above specification is not very efficient in that it makes frequent use of the path planner, which is relatively expensive. We would like to devise some heuristics to restrict these calls.

Note that if a pallet can be collected, there has to be enough space around one of its picking-up sides for the AGV to approach. If we use free(X, Y) to mean that X has a free side in the existence of the pallet set Y and other obstacles and redefine `sortbypath` as

```
sortbypath([]).
sortbypath([H|T]):-
        free(H,T),
        path(H,T),
        sortbypath(T).
```

then the path(A, B) test will be carried out only when the pallet A has a free side and therefore the total number of calls to `path` will be reduced.

Another heuristic we use is the *nearest pallet heuristic*—always testing the pallet that is nearest to the goal (the processing centre) first of all. This comes from the observation that the nearer a pallet, the better chance it has not to be blocked by other pallets.

12.3 Initial Path Planning

The ability to plan collision-free paths is critical in the solution of our problem. Our implementation has access to two built-in path planners, and in increasing order of competence these are one based on a visibility graph (Chapter 10) and the other on a Voronoi diagram technique in C-space (see §12.5). Both planners are generic, in that they are not tailored to a particular vehicle, but they are useful in *quickly* generating

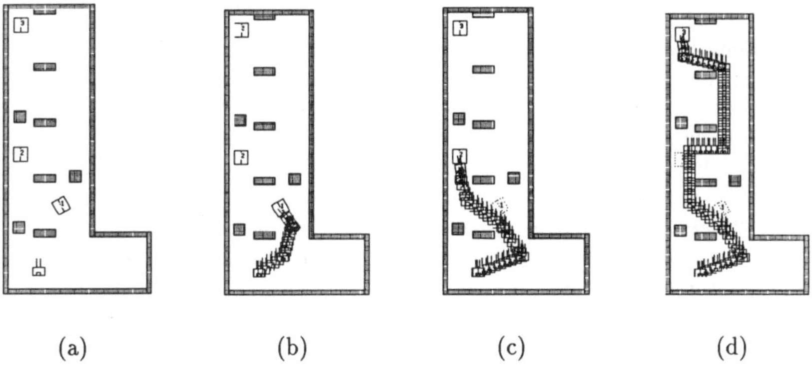

Figure 12.3: A collection process of the AGV in the AGV lab.

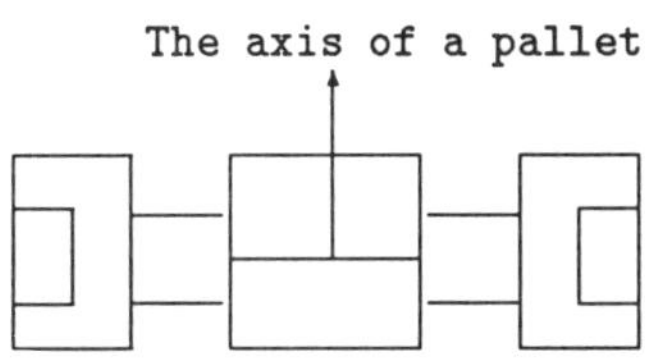

Figure 12.4: The axis of a pallet and the goal positions

potential paths, and in rejecting infeasible paths. We then call the 'exact' planner for the particular vehicle (Chapter 11) to refine the remaining paths.

An example of a visibility graph based plan can be found in figure 12.3. For simplicity, the C-space based planner assumes that the vehicle has only two translational degrees-of-freedom. Recall that the uncollected pallets may also block the way of the AGV so they also affect the configuration space. The configuration space thus changes due to the removal of pallets during the collecting process. A visibility graph for the C-space is built *lazily*, *i.e.* an edge is added only when necessary. As what we are concerned with are those nodes that are reachable from the starting position, we begin with the starting point and extend the graph with reachable nodes until the goal is included or no more nodes can be included.

We allow partial solutions under certain circumstances when some of the pallets cannot be collected. The program will detect these cases and give the order of those that can be collected.

The 'get into touch" problem poses additional constraints on the AGV's motion.

For example, the AGV has to approach a pallet in one of a small number of orientations, say along the axis of the pallet (see figure 12.4). Instead of building these features into the geometric path planner to make it less general, we isolate the extra constraints and solve them independently. The path of the AGV when approaching a pallet is hence split into two segments: one takes the AGV from the processing centre to some position around the pallet and the other brings the AGV into contact with the pallet. The first segment is planned by the external real time planner. After taking this segment, the AGV should align itself with the axis of the pallet and get its forklift against one picking-up side of the pallet. The palletiser thus extracts the information about the pallet from the world model, calculates its axis and position, and passes these parameters over to the path planner as the new subgoal.

12.4 Refining the Scheduler

The scheduler we outlined above produces a linear ordering of the pallets for the AGV to collect, according to data held within the world model at plan time. However, the world model may not be entirely accurate: at run time the AGV may find some unexpected obstacles, and replanning will be necessary [44]. The scheduler should thus be *robust* to minimise replanning, *i.e.* to provide immediate alternatives when the AGV fails to access the scheduled goal.

To cope with these problems, we use an approach based on *scheduling graphs*. We define a scheduling graph to be an undirected graph specifying certain relations between pallets, as inferred from the world model. All the possible collecting orders are encoded within the graph and a total ordering can be easily derived by traversing the graph. When the AGV's run-time access to a goal fails, alternatives are readily picked out from the graph.

> **Definition:** A *scheduling graph* is an undirected graph in which the nodes denote initial configurations for the pallets and the robot, and two nodes are connected if there is a path for the robot plus pallet from one node to the other.

An environment and its scheduling graph are shown in figure 12.5.

This approach to scheduling follows that of Fox and Kempf [71]. They discussed the problems of robotic assembly and suggested enabling the robot to "reason *opportunistically* over knowledge of part availability at run time". They modelled the temporal constraints as partial order relations and proposed 'variable orders' to be derived upon these relations to cope with the uncertainty of part availability.

The idea of scheduling graphs are similar to that of Fox and Kempf's 'sets of partial order' scheme but, in our case, partial order relations are not representatively powerful enough. Although, as Fox and Kempf pointed out, the graphs are also incapable of representing all kinds of constraints encountered in scheduling problems [71], they

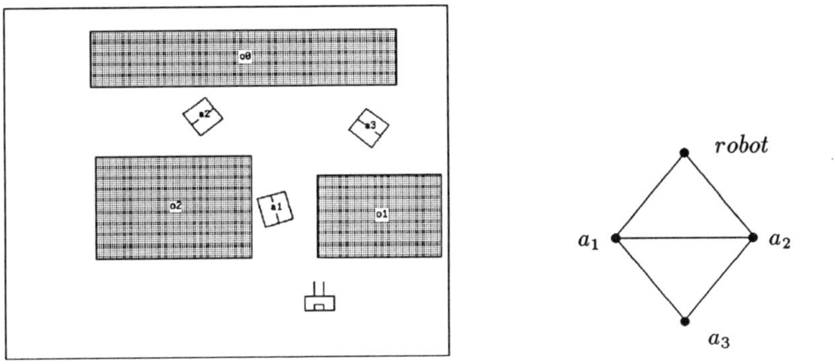

(a) The environment (b) The scheduling graph

Figure 12.5: An example of a scheduling graph

work well for our problem as our problem will not involve 'negative' constraints, such as "task a cannot be the last", which cause trouble.

The use of scheduling graphs to derive collecting orders can be specified by a PROLOG program. In practice, the scheduling graph could be calculated beforehand and dynamically maintained at runtime. If the current goal is not reachable (due to unexpected obstacles), the path planner can then notify the system which will subsequently schedule the AGV to a nearest alternative goal (node); otherwise it removes the node from the graph and links all those pairs of nodes that were both linked to the removed node previously. The same procedure is repeated for the new graph, and the process continues until there are no more pallets left or none of the remaining pallets are accessible.

12.4.1 Voronoi Scheduling Graphs

The scheduling graph encodes all the orders that are geometrically possible, but takes no account of the degree of difficulty associated with executing a path; but in figure 12.6, the AGV will have difficulty in collecting a_4 before a_3. Thus it is desirable to refine the path relation, and for this we use *General Voronoi diagrams* (GVDs) (see Chapter 10).

Given an environment E, the GVD is defined with respect to the obstacles and boundaries. The GVD itself has a graph structure, in which the edges are points in space which are equidistant from two objects, and the vertices are equidistant from three or more objects. The regions between the objects and the GVD proper thus partition the space that is nearer to one object than any other, and we call these the

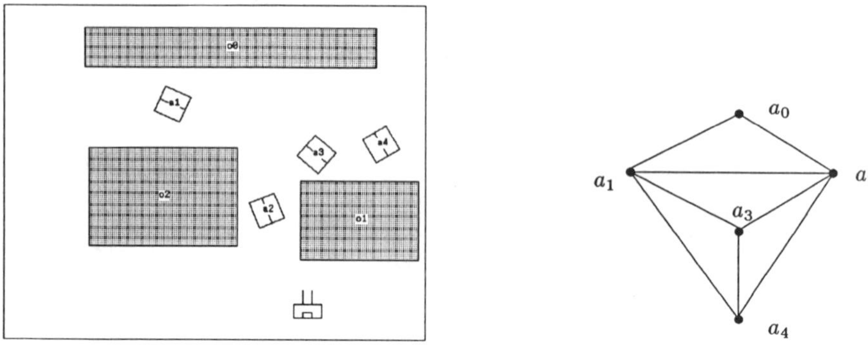

(a) The environment (b) The scheduling graph

Figure 12.6: An example of awkward orders (See text for detail)

Voronoi regions. A pallet will generally lie within some Voronoi region, and we can draw a line (or small number of lines) of minimum length from the centre of the pallet to the GVD. If we call the points where these lines meet the GVD the *v-image(s)* of the pallet, then a reasonable path for the pallet to go from a starting configuration to a goal configuration is for it to move onto the GVD (i.e., to a v-image), then along the GVD to a v-image for the goal, and thence to the goal. (This idea will be refined in §12.5.)

> **Definition:** A *Voronoi Scheduling Graph* is a refinement of a scheduling graph with an additional constraint on the placement of arcs, namely that there is a path along the GVD between the v-images of the two pallets which does not cross the v-image of any other pallet.

Voronoi scheduling graphs limit the number of the possible orders that can be derived from scheduling graphs, by removing paths that would cause a pallet to pass 'too close' to another pallet, and thus produce more reasonable ordering for the AGV. In figure 12.7, we show the GVD of the environment of figure 12.6 and the corresponding Voronoi scheduling graph. An order derived from this graph, for example $a_1a_2a_3a_4$, will be easy to follow.

12.5 A GVD-based Path Planner for Mobile Robots

We use within the system a second internal path planner for mobile robots based on GVDs. Ground-based mobile robots, although themselves three dimensional rigid

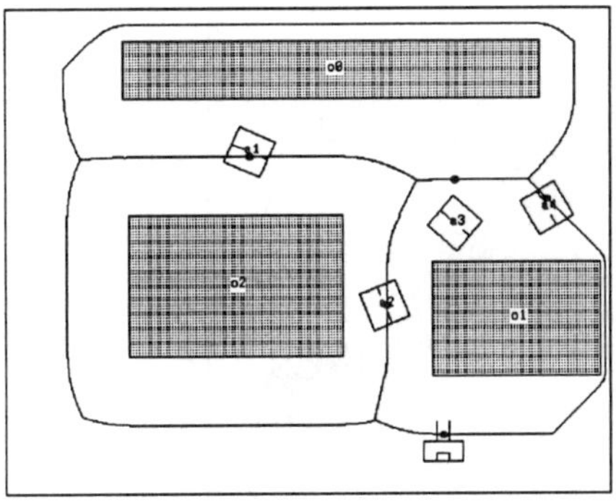

(a) The environment and GVD

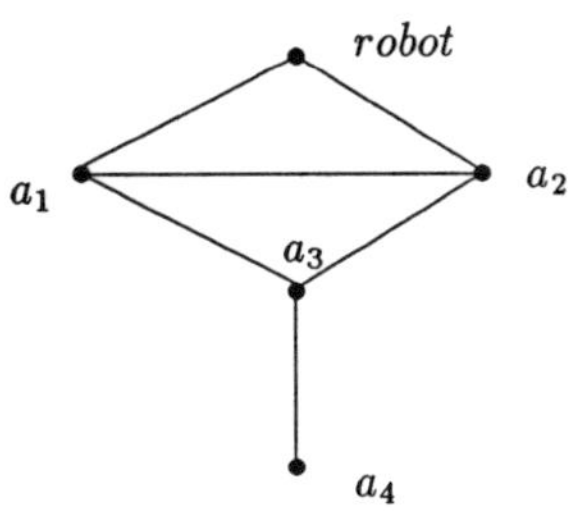

(b) The Voronoi scheduling graph

Figure 12.7: A Voronoi scheduling graph

objects, have only three degrees of freedom: two translational and one rotational, all on the ground. If, as is often the case, their cross sections do not change much with the height, it is realistic and reasonable to project the environment onto the moving plane and reason using this projection. This reduces the three-dimensional planning problem into a two dimensional one. Such a projection we call a 'map'. A second feature of mobile robots is that they can usually be boxed by rectangular hulls as their length and width do not differ greatly. And finally, unlike multi-link manipulators, mobile robots often work in reasonably open environments. However, there is a need for them to keep a reasonable distance away from obstacles, as the control of such vehicles is normally not precise.

The generalised Voronoi diagram has been used by many researchers to solve path planning problems [152, 104, 187]. The reason why they are so popular is twofold. On the one hand, a very important criteria to judge a good path is that it is smooth (so that the final trajectory generated from it can be smooth). The GVDs are networks of linear segments with parabolic blends which are smooth except at the Voronoi nodes. On the other hand, GVDs are the locus of points that are equidistant from at least two obstacles. Following them, robots will tend to move in a way that keeps them away from obstacles as far as possible.

However, letting the robot purely move along the GVD with its reference point sticking to it, as many existing GVD schemes do, does not produce good paths. This is partially because every Voronoi edge is composed of many pieces which, although continuous, exhibits a lot of turns. By following it, the robot will move in zigzags even when a simpler path may exist. The second reason is that GVDs are not smooth at

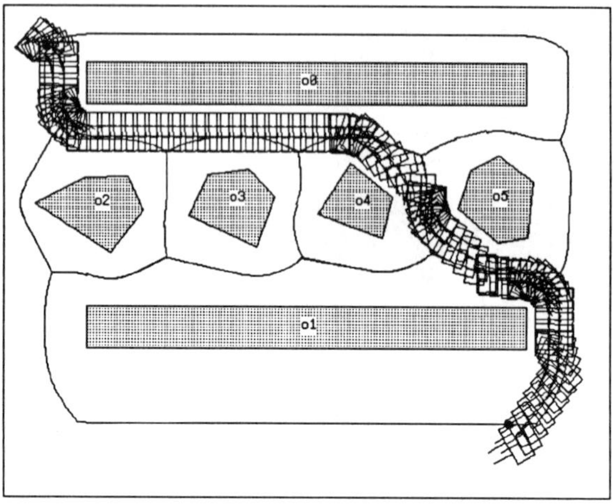

Figure 12.8: An example of the GVD path

Voronoi nodes, which are usually used as junctions when planning a path. This may force the robot to take sharp turns at some junctions. Finally, when the workspace is sparsely littered with obstacles, the GVD path can be *over robust*, *i.e.* it may keep unnecessarily far away from some obstacles, clearly a waste.

The planner we implemented is not meant to investigate new heuristics to guide the robot through narrow gaps. Instead, and with the above problems in mind, it is an attempt to produce more practical paths for robots to follow. The approach combines a cell decomposition method with a retraction method. However, the cells in our context are regions of the free space rather than those of the configuration space. The idea is to first find those cells that the reference point of the robot will go through, and then plan locally in each cell. Such a planning process in practice is very similar to a process where a GVD global path is first derived and then refined to a 'good path'. But by use of cells as an intermediate step, it is made clearer how such a refinement procedure is conducted. The cells are derived from the Voronoi regions.

Briefly speaking, the algorithm works in the following way: first the Voronoi diagram of the environment is calculated and the free space decomposed into cells. Then, we search for a global path which, unlike other methods, takes the form of a sequence of cells. Finally, local paths are planned within each cell that the robot will go through, and these are linked across the cells. The orientations of the robot are specified at this stage as well. An example of a computed path is shown in figure 12.8.

We use the following conventions:

1. the robot is covered by a rectangle and the path is planned with respect to this shape. The reference point is chosen as the centre of the rectangle.

2. The workspace is bounded by a rectangular box, which is regarded as an unbounded obstacle, and every other obstacle is a convex polygon.

3. When we talk about the position of the robot, we mean the position of its reference point. A configuration of the robot is its position associated with an orientation.

4. For the path to be robust, a minimum clearance of `minclr` is to be kept. Without losing generality, we take `minclr` to be the minimum distance between the reference point and the obstacles. Suppose that the robot's rectangular hull has dimensionals h by w. A necessary condition for `minclr` is `minclr` $> w/2$. But in this case, the robot may touch obstacles when it turns around a corner [187]. The heuristic we take is `minclr` $> ((h/2)^2 + (w/2)^2))^{1/2}$, *i.e.* the global path can be traversed by a moving circle whose diameter equals to the diagonal of the robot in the free space. This is consistent with our assumptions that the mobile robots are not very elongated and work in less cluttered environments (and therefore the 'channels' between the obstacles are reasonably wide). In doing so, the global path found will be wide enough for the robot to take a desired orientation at any point on the path. It also allows a degree of tolerance for the uncertainty.

To search for the global path, we first build a weighted graph derived directly from the Voronoi diagram. Then, we search this graph for a path and refine it into the cell sequence we are after. Having divided the space into cells using the GVD, and figured out which cells the robot (*i.e.* its reference point) should go through, we plan locally a piece of path within each cell, and link those pieces across the cells.

The cell sequence can be broken into consecutive segments. Each segment consists of the cells belonging to the same Voronoi region. The local paths within the cells of a segment are hence computed with respect to the same obstacle. We can merge these local paths into one single local path. Therefore, the final path will be composed of local paths within the Voronoi regions and *slip ways* to link the local paths across the regions.

The basic strategy to plan a local path is to keep the robot away from the obstacle by a distance of `minclr`. The local path thus consists of straight line segments along the edges of the obstacle and circular blends of `minclr` radii around the vertices of the obstacle. The interchanges of the straight line motion and circular motion happen when the robot meets a normal ray (or *N-point*, see figure 12.9). As the straight line is tangent to the circular segment, the local path is a smooth curve.

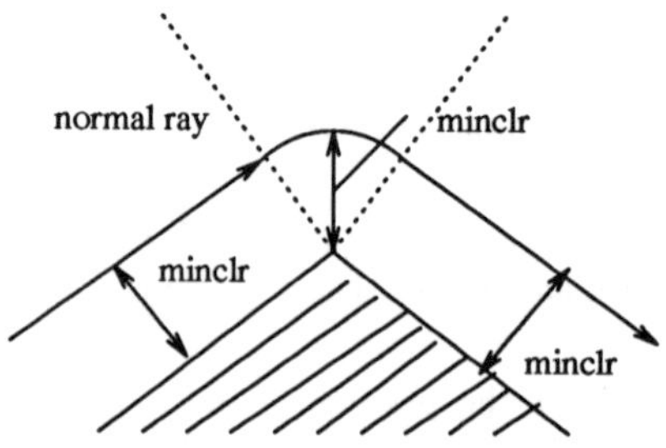

Figure 12.9: The local path along an obstacle

12.5.1 Heuristics to Smooth the Linkage

Following exactly a path designed by the above method, the robot sometimes has to take a sharp turn away from the local path to the slip way or vice versa. At these situations we attempt to apply a smoothing heuristic, by which we let the robot leave (start respectively) the local path earlier (later, respectively) to take a smooth slip path to (off, respectively) the N-point. An example is shown in figure 12.10. Without the heuristic, the slip path leading to the local path of o_1's Voronoi region is almost normal to its previous moving direction, as shown on the left. It is improved when the heuristic applies, as shown on the right. A second heuristic prevents the robot from unnecessarily pulling-in to an obstacle's edge and then pulling-out again.

The second heuristic can be extended to cope with another pulling-in and pulling-out situation. This time the robot unnecessarily pulls in to an obstacle and then pulls out again. This usually happens in a less cluttered environment. Consider the example shown in figure 12.11. The robot is first scheduled to go through the regions of $\cdots o_1 o_0 o_2 \cdots$ where o_1 and o_2 are neighbours. The Voronoi edge defined by o_0, o_1 and o_0, o_2 have rather large clearances. Taking a local path within the Voronoi region of o_0 in this case is obviously a waste. Thus, the points on these Voronoi edges should no longer be the via points. The heuristic moves the appropriate via points *normally* towards the obstacle o_1 or o_2 until they are away from the obstacle with a distance of `minclr`. Now if the robot can take a straight line collision-free motion between these new N-points (this can be checked using a collision-detection technique [43]) then it is the original path we are after; otherwise a local path is taken within each region towards each other until at some point every local path can see each other.

The last heuristic is especially useful when the robot needs to go through cells defined by the boundary of the workspace according to the computed global path. It allows the robot, instead of moving towards the boundary, which is often a waste, to keep well inside the workspace.

Lastly, the robot keeps an orientation that is tangential to the path when taking a local path. This is guaranteed to be continuous as each local path is smooth. However, it is highly possible that continuity is lost at the slip ways. In this case, we let the

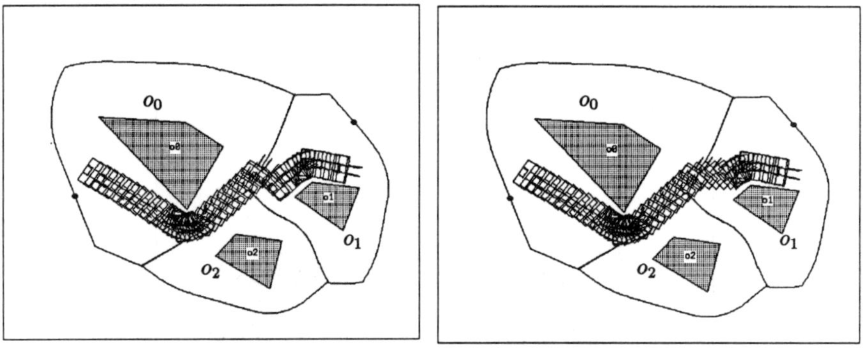

(a) Without the heuristic. (b) With the heuristic.

Figure 12.10: The sharp-turn heuristic

robot rotate while translating to get itself aligned with the supposed orientation at the beginning of the next local path[1].

12.6 Putting It All Together

Given an environment with the pallet set $A = \{a_0, a_1, \ldots, a_n\}$ and obstacle set OBS, we attacked the problem as a repetitive collision-free path checking problem in a constantly changing environment (due to the removal of pallets), and developed the specification of the problem as given in §12.2. As the robot can collect the pallet from one of the (two) picking-up sides, if we define c_findpath($P, si, Prest$) as a geometric primitive which returns whether there exists a path for the robot to collect the pallet P on side si in the presence of the remaining pallets $Prest$ and other obstacles, then the PROLOG predicate path can be further refined into

```
path(P1,Prest):- append(Prest,OBS,All), path01(P1, All).
path(P1,Prest):- append(Prest,OBS,All), path23(P1, All).
path01(P1, All):- X is c_findpath(P1,'01',All), X = yes.
path23(P1, All):- X is c_findpath(P1,'23',All), X = yes.
```

[1] All of this may seem a little *ad hoc*, but recall that the purpose is only to find reasonable paths for a fairly generic vehicle: the planner of Chapter 11 can later be used to produce good paths.

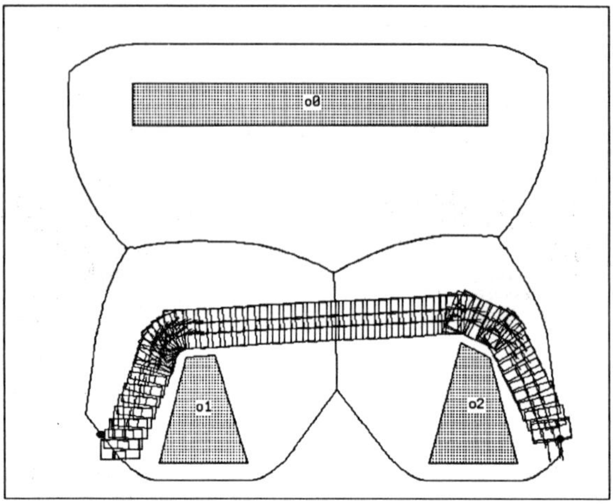

Figure 12.11: The pulling-in and -out heuristic for obstacles

12.6.1 Improving the Solution by Heuristics

As the full path planner is expensive, we have introduced some heuristics into the planner to reduce the total number of calls to the path planner. This can be done easily at this higher-level.

The Nearest Pallet Heuristics

Among all the pallets, the nearest one to the goal has a better chance of not being blocked by other pallets, and hence has a better chance of being collected. Therefore, at any step of planning, we always choose the nearest pallet first for path existence testing. If it cannot be collected, the second nearest will be chosen, and so on. This is equivalent to sorting the pallets according to their distance to the goal before planning.

We define the geometric primitive `c_dist`(X) to be the distance between pallet X and the goal. An order `ordered` is assigned according to `c_dist`:

```
ordered(B,A) :-
        X is c_dist(B),
        Y is c_dist(A),
        X < Y.
```

The solution can be improved into

```
findorder(Original,Result):-
```

```
            sortbydis(Original, Sortbyd),
            permutation(Sortbyd,Result),
            sortbypath(Result).
```

where `sortbydis`(X,Y) sorted X into Y according to the order defined above.

R-dependency Relations

If a pallet can be collected, there has to be enough space around its picking-up sides for the robot to approach. We earlier formulated this relation by the predicate `free`, and in turn `free` can be refined by the following R-dependency relations.

> **Definition:** A pallet a_i is said to be *R-dependent* on the pallet a_j if, when combined with the robot on its picking-up sides, a_i overlaps with a_j. In particular, a_i is R-dependent on `OBS` (for obstacle) if the combined body overlaps with any obstacle or bumps into walls.

As every pallet has two picking-up sides (side 01 and side 23), the R-dependency can be refined into two relations: `R_dep01` and `R_dep23`. In figure 12.12, dotted lines outline the robot combined on the picking-up sides of pallets. We have `R_dep23`(A, B), `R_dep01`(B, A), `R_dep01`(D, C), `R_dep23`$(B,$ `OBS`$)$, `R_dep23`$(D,$ `OBS`$)$.

Thus a pallet is free if it is not R-dependent on any other pallet or obstacle on a picking-up side. In figure 12.12, for example, we have `free`$(A, [B, C, D])$, `free`$(B, [C, D])$, `free`$(C, [A, B, D])$, `free`$(D, [A, B])$, ...

We define a geometric primitive `c_rdep`$(P1, P2, si)$, which returns true if the pallet $P1$ is R-dependent on the pallet (or obstacle) $P2$ on the side si. Then the program can be improved by the following redefinition of `sortbypath`:

```
sortbypath([]).
sortbypath([P1|Prest]):-
        free01(P1,Prest),
        path01(P1,Prest),
        !,
        sortbypath(Prest).
sortbypath([P1|Prest]):-
        free23(P1,Prest),
        path23(P1,Prest),
        !,
        sortbypath(Prest).

free01(P1,[]) :- not(rdep01(P1,OBS)).
free01(P1,[H|T]) :- rdep01(P1,H),!,fail.
free01(P1,[H|T]) :- free01(P1,T).
free23(P1,[]) :- not(rdep23(P1,OBS)).
```

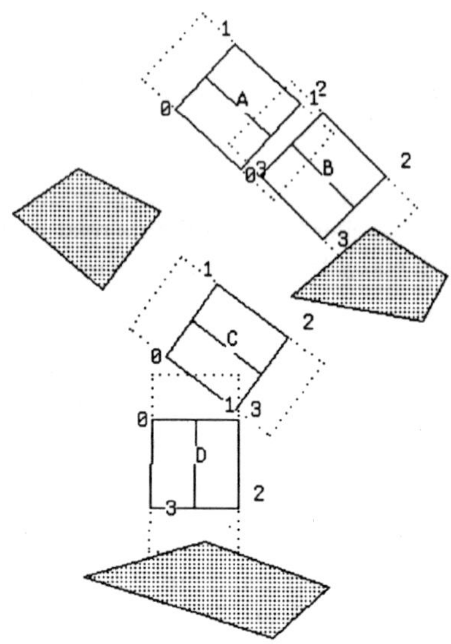

Figure 12.12: The R-dependency relations

```
free23(P1,[H|T]):- rdep23(P1,H),!,fail.
free23(P1,[H|T]):- free23(P1,T).

rdep01(P1,P2):- X is c_rdep(P1,P2,0), X = 1.
rdep23(P1,P2):- X is c_rdep(P1,P2,1), X = 1.
```

The remaining primitive relations are concerned with the geometry of the objects and can be more efficiently implemented in a conventional language; we used the language C, which can be directly called under the POPLOG environment.

12.6.2 Optimisations

The path checker is usually expensive and every effort should be taken to reduce the frequency of invoking it. We have already introduced some heuristics to restrict the number of such calls. Here we extend the optimisation by retaining some history information.

We maintain two set of predicates in the data base. One set is of the form paths(p, si, S) indicating that there is a path for the robot to collect p on side si

amongst the set of pallets S and other obstacles. The other is **nopaths**(p, si, S) indicating the opposite. In both cases, we implement S as a set (rather than a list). Each clause is added to the knowledge base as a result of the corresponding call to the path checker; and every time before we make an call to the path checker, we check the database to see whether there is an appropriate clause indicating the existence or non-existence of the path. If there is, redundant calls can be avoided.

Moreover, we observe the following laws for **paths** and **nopaths**:

$$\begin{aligned} \texttt{paths}(p, si, S) \wedge (S_1 \subseteq S) &\implies \texttt{paths}(p, si, S_1) \\ \texttt{nopaths}(p, si, S) \wedge (S \subseteq S_1) &\implies \texttt{nopaths}(p, si, S_1) \end{aligned}$$

The first indicates the fact that if the robot can find a path to approach a pallet p despite the existence of the pallet set S, then it must be able to find a path to p despite the existence of any subset of S; while the second indicates that if the robot cannot find any path to approach p due to the existence of pallet set S, then it must not be able to find a path to p due to the existence of any superset of S.

As the role of the path checker is to check the existence of paths rather than produce accurate paths for robots, the application of these laws allows us to make more efficient use of the predicates **paths** and **nopaths**, and to further reduce the number of calls to the path checker. To code them to the system, we modify the definition of `path01` into the following:

```
path01(P1, Prest):-
        paths(P1,'01',Others),
        subset(Prest,Others).
path01(P1, Prest):-
        nopaths(P1,'01',Others),
        subset(Others,Prest),
        !,
        fail.
path01(P1, Prest):-
        X is c_findpath(P1,'01',Prest),
        X = yes,
        !,
        asserta(paths(P1,'01',Prest)).
path01(P1, Prest):-
        asserta(nopaths(P1,'01',Prest)),
        fail.
```

where `subset`(A, B) indicates that A is a subset of B.

`path23` is redefined similarly. Note that the same technique—of caching useful results—could also be used for `free01` and `free23`.

12.6.3 An Example of Running the System

An example is shown in Fig. 12.13. The indices of the pallets are indicated on the pallets. Running the program with this environment, the pallet list is firstly sorted according to their distance to the goal:

```
[3, 2, 0, 1].
```

The R-dependency relation is then calculated:

```
R_dep01(0,OBS).
R_dep01(1,OBS).
R_dep01(2,3).
R_dep23(2,OBS).
R_dep23(3,2).
```

where `OBS` stands for obstacles and walls as a whole.

The system called the path checker a total of 9 times and the results are recorded in the system's database for reuse:

```
paths(1,23,[3, 2]).
paths(0,23,[3, 2, 1]).
paths(1,23,[2]).
paths(0,23,[2, 1]).
paths(1,23,[]).
paths(0,23,[1]).
paths(2,01,[0, 1]).
paths(3,01,[2, 0, 1]).

nopaths(1,23,[0]).
```

The backtracking control of PROLOG allows the system to have multiple solutions, and the derived orders are:

```
[3, 2, 0, 1];
[3, 0, 2, 1];
[3, 0, 1, 2];
[0, 3, 2, 1];
[0, 3, 1, 2];
[0, 1, 3, 2].
```

PROLOG programs usually work in a generate-and-test way, and this could cause redundant computations unless the program is carefully written. For example, even if the system has already calculated that the order `[1, 2, 3, 0]` is impossible because there is no path to collect pallet 1, it will still generate the permutation `[1, 3, 0, 2]` and test it. But by retaining facts in the database, this redundancy can be avoided.

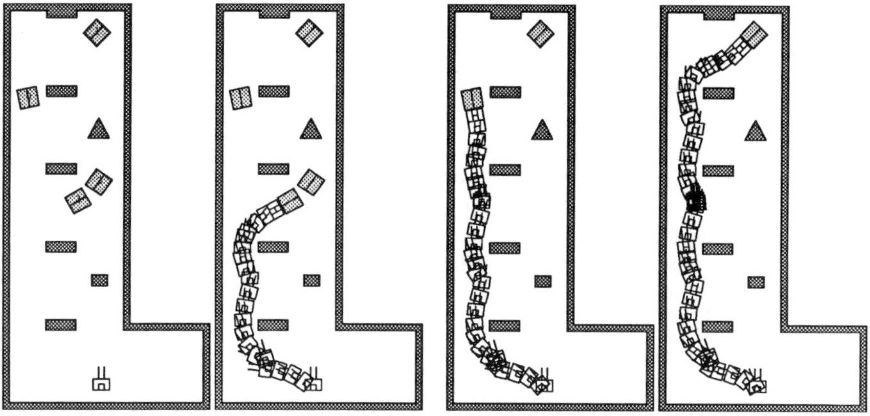

(a) (b) (c) (d)

Figure 12.13: Potential field paths for collecting pallets

12.6.4 Linking with External Path Planners

The C-space planner embedded in the system is simple yet fast. It is used to check the potential existence of paths. Such checks are frequent and therefore the speed of this planner has a great effect on the speed of the overall planner. To generate the final trajectory, we called upon the more powerful planner described in Chapter 11.

The hierarchical structure of our system certainly helps us to incorporate the independent external path planner into the system. The planner is coded in POP11 (but it doesn't matter if it is coded in another language). To use the planner, we set `environment` to be the initial working space, `vehicle` to be the robot we are using and `start` to be the pallet processing centre. Then, the system will suggest an initial pallet to be collected, and call upon the path planner to try to collect it. Afterwards, `environment` is modified by removing the pallet already collected and a new target pallet selected. Such a process repeats until all the pallets are collected.

12.7 Remarks

We have developed a hierarchical robot reasoning system, with different levels employing different knowledge representations in order to make explicit the knowledge represented. In doing so, we allow the strategies and heuristics to be coded in a higher-level representation which, in turn, enables them to be easily replaced or improved. Also, building the system in this way allows the system to be extensible and easily adapted to changes, such as the incorporation of new constraints.

The final planner consists of a geometric layer of around 10 000 lines of C code, plus a connection to the external path planner. This layer is difficult to write, and thus difficult to change. However the *strategy* in the planner is entirely encoded within the PROLOG code. This consists of a total of 50 lines, and is easily changed, both in order to change the problem strategy, and indeed to change the problem. The use of the geometric primitives allows the rule-based engine to be separated from the geometric data, which would otherwise make the rule base facts more complicated. Moreover, the separation of the rule-based engine from the geometric experts allows them to be developed in different modules. The changes made upon one of them will not influence the other.

We view the finished system not as a perfect palletising planner, but rather as a prototype for solving a number of problems involving motion in the plane. For example, it should be possible to *easily* change the task that the system is trying to perform, and indeed we have demonstrated this with the system by getting it to perform a different task, namely *sofa moving.* The goal here is to move a sofa from a start position to a goal position, using two robots as shown in Fig. 12.14. The two robots are assumed to be able to move arbitrarily, and to be able to move the sofa arbitrarily by both getting to one end of the sofa, and lifting it up (as in Fig. 12.14(a)). Alternatively, the sofa can also be pushed, using just one robot, but only in the direction of its long axis. A final complication is the presence of small obstacles that are not known at the initial plan time, but can be detected as the robots near them at run time, and can be pushed out of the way by a robot (as in Fig. 12.14(b)). We do not intend to give detailed results of this experiment here, but note that the prototype system was re-implemented to solve this problem in about two weeks. The final PROLOG code also runs to around 50 lines, and much of the development time was taken up with cosmetic changes to the interface to the geometric layer of code. This work has also confirmed to us that the level of communication between the strategic planning level and the geometric planning level is about right; the communications at this level involve concepts such as distances, (indexes to) pieces of path and GVDs, and requests to check for collisions and to generate sub-goals.

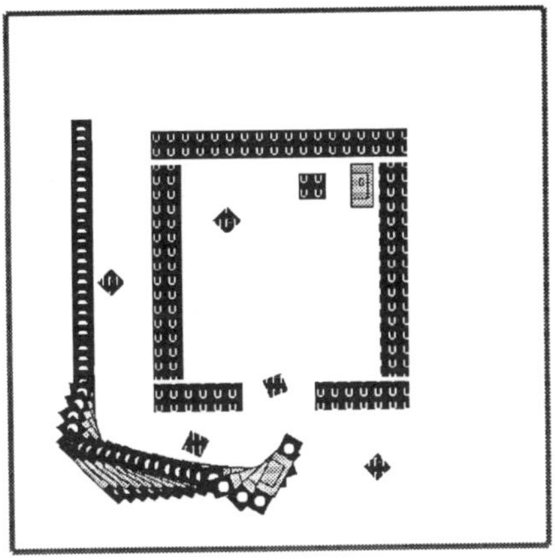

(a) approaching an obstacle

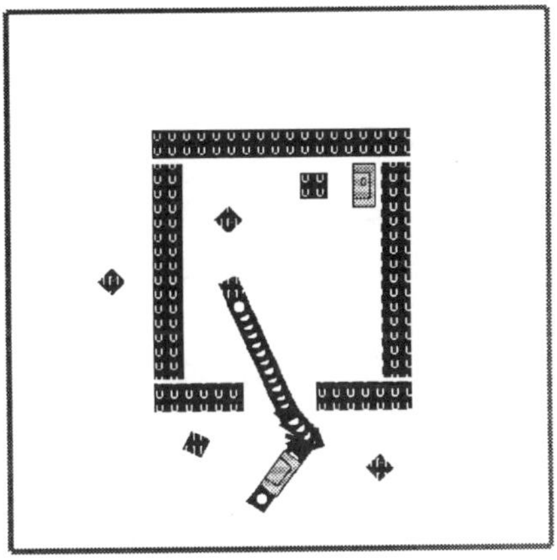

(b) pushing the obstacle away

Figure 12.14: The sofa-mover in action

Chapter 13

Modelling

Bill Triggs, Tony Hague and Stephen Cameron

The last few chapters have introduced several applications that rely on the direct use of geometric data-structures, rather than sensed information. For the application areas addressed in the initial phases of the AGV project it is reasonable to expect that a CAD database of the major structures in the vehicle's environment will be available, and so it makes sense to be able to use that information, rather then only relying on sensed information. For the AGV project we made a deliberate decision to pay considerable attention to the storage of geometric data. Previous work on geometric planning has generally used *ad hoc* data-structures, as the researchers have been more concerned with the problems of planning than adherence to standards. In our case, although we were not trying to develop production code, we were concerned that our techniques should scale up to factory-sized models. Just to give an idea of what that involves: all of our planners use two-dimensional polygons as data, and in the laboratory reasonable models of the environment require tens to hundreds of polygon edges. However a factory model will easily contain tens of thousands to hundreds of thousands of polygon edges. At this sort of scale we need to be able to access and update all or part of the geometric data—or *map*—quickly.. A hierarchical data-structure is thus used (rather than a linear data-structure). Further, a real factory system will employ many vehicles, and so changes to the data-structures need to be distributed amongst several computational agents. This is a thorny problem which is just starting to be addressed by the distributed database community.

This chapter therefore starts by describing the requirements of and design of the map for the AGV, called the *world model*. We then describe a particular geometric computation that is designed to run on top of the world model, namely that of extracting 'natural corridors' from a model to act as roadways for the path planning algorithm of Chapter 11.

13.1 Requirements

The AGV has a number of advanced sensor systems for navigation, obstacle avoidance and pallet acquisition, and a corresponding set of task and path planning strategies, but at the heart of the software is a model-based representation of the vehicle and its environment, supported by a purpose-built geometric database.

There are several good reasons for providing a robot with such a world model:

1. It provides an effective, easy to use, high level interface to the robot: task level behaviour such as 'pick up the pallet at x and carry it to y' is natural to a robot which 'understands' its surroundings.

2. It encourages a correspondingly clean internal architecture in which independent vehicle subsystems such as sensing and data fusion, vehicle controlling, and path and task planning modules communicate with one another mostly via the world model. (Albeit there are usually a few time or safety critical mechanisms such as basic vehicle reflexes which must bypass the model). This modular, inherently parallelisable, model-based, data-driven architecture simplifies the introduction of new subsystems by providing a uniform inter-module interface; in particular it allows vehicle capabilities to be extended by the addition of more or less generic 'expertise modules' which need only interface with the relatively simple and vehicle-independent model database.

3. It decouples essentially robot-independent environmental information from the robot-specific software, allowing it to be handled by efficient specialist code such as the spatial database and shared between a number of robot and control processes. The result is an integrated model of the whole environment which is immediately extensible to large multi-robot systems and provides a solid foundation on which to build interacting-robot and factory-wide controllers.

Fig. 13.1 illustrates the model-based architecture we designed for the AGV project. Before discussing this in detail we shall briefly list some of the 'expert modules' that have been developed by other contributors to the project:

1. The Constructive Solid Geometry engine ROBMOD [45, 47], which was written by Stephen Cameron, was used for all of the solid modelling on the project.

2. The main path planner for the vehicle (Chapter 11) is designed to use fixed geometric data (rather than the sensed data employed in Chapter 4).

3. An 'expert palletiser' was developed (see Chapter 12 and Chapter 9) to plan the task of clearing a region cluttered with pallets and obstacles. This requires

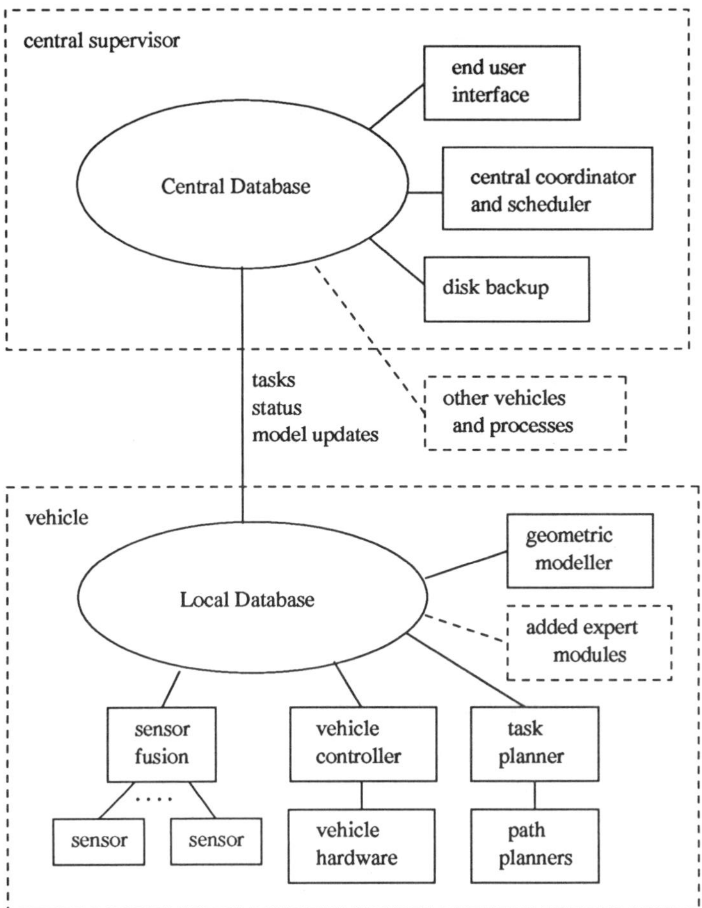

Figure 13.1: A modular architecture for a world-model based multi-robot system

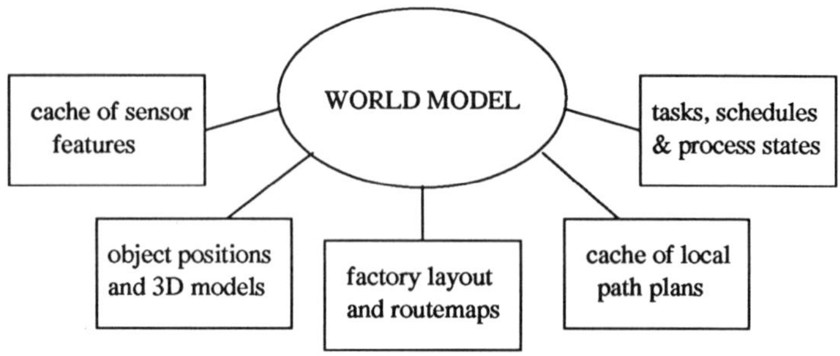

Figure 13.2: Types of information stored in the robot world model

the integration of the above path planner, a rule-based system enhanced with several graph-based heuristics to select a suitable order for clearing the pallets, and a docking manœuvre planner to supervise the vehicle as it approaches and picks up pallets.

All of these modules are designed to be sufficiently flexible to be used with other vehicles or systems.

A useful robot world model must contain a wide variety of information, for example (see Fig. 13.2):

1. Representations of the physical layout of the environment and the objects within it such as walls, pillars, pallets and robot vehicles.

2. More detailed descriptions of objects such as three dimensional solid or boundary models.

3. Maps of features important to model-based sensor systems, such as visible edges and sonar-reflecting walls and corners.

4. A decomposition of the workspace into functional regions such as roadways, processing sites and stores, and a corresponding route-map graph for large scale route planning.

5. Caches of reusable local or temporary data such as details of paths planned around local obstacles and partial feature maps of unidentified objects.

6. State, task, and scheduling information for the modelled robots and manufacturing processes.

The critical component in this model-based architecture is the database which underlies the world model: it must be able to store and access at least the above types of information; it must be particularly efficient for the common geometrically or spatially organised data; it must interface easily both with the end user and with a potentially wide range of robot subsystems; and it must be sufficiently flexible to deal with unexpected types of information and subsystem as neither of these are known perfectly in advance.

Moreover, even for a single robot system such as the AGV we feel that the model database should be capable of supporting a distributed multi-robot multi-process world model because:

1. This allows the robot software to be simply partitioned into a number of independent modules, each with access to the database, as above.

2. Even a robot as autonomous as our AGV must exchange a considerable amount of task, state and environmental information with a land-based supervising computer over a (low-bandwidth radio) link, so that both vehicle and supervisor world models of some sort are required, together with a reasonably efficient data-exchange protocol. It seems sensible to use the same format for both model databases and organise the communications around database records, in other words to build a primitive distributed database system: this distribution may as well be built in to the database at the outset.

3. A robot designed around a multi-robot database is more easily adapted to an integrated multi-robot environment. Such integrated systems are interesting in their own right and are likely to be of very considerable practical importance in the future.

The support of such a multi-process system requires some sophistication of the database, in particular the consistency of multiple distributed copies of database records must be maintained in real-time, preferably without undue complication to either the user interface or the implementation.

13.2 The Oxford Robot Database

We developed our own modular object-oriented database system specifically for robot modelling in a distributed environment.

13.2.1 Design

In this system generic header structures known as 'Objects' are attached to all data items; for the most part the database uses only these header fields, however one such field describes the type of data contained in the Object and if necessary the database can manipulate this private data with the type-specific Object routines supplied with each type. The Object header is intended to provide a succinct description of the enclosed data suitable for use with a range of indexing systems, however it actually contains enough information to describe simple physical objects in its own right (see Fig. 13.3). Fields are provided for an object name string and a unique system-wide

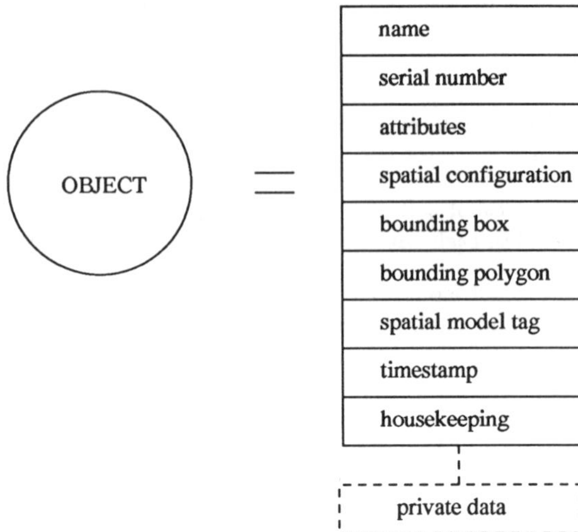

Figure 13.3: The Object header structure

object serial number, for a vector of user-defined binary 'object attributes' and a time-of-last-modification of the object, and most importantly for a spatial configuration (position and orientation) and a description of the spatial region occupied by the object. The main types of index currently supported correspond to these fields — in particular it is possible to access an object by name or serial number and by two or three dimensional spatial position — however new types of index are very easily added.

The spatial parts of the system are 'two and a half dimensional': all positions are three dimensional but coordinate transformations always preserve the vertical axis so that object orientation is given by a single angle. The spatial extent of an object is specified by a three dimensional bounding box (aligned with the current coordinate system) and an optional two dimensional inscribed polygon; these allow the database to perform fast but fairly crude object localisation and interference detection, and efficient spatial indexing. The database kernel deliberately does not have solid modelling expertise, since if full three dimensional shape descriptions or interference tests are required (for example for visible edge prediction or path planning in a cluttered region) a specialist geometric modeller can be (and is) easily run as a database subprocess. (There is a tag in the Object header for a modeller-based shape description). At present this role of 'system geometry expert' is filled by the Constructive Solid Geometry modeller ROBMOD [45, 47], however the modeller is not tightly coupled to the database and could easily be changed or omitted if required.

In operation, the database kernel attempts to maintain appropriate multiple copies

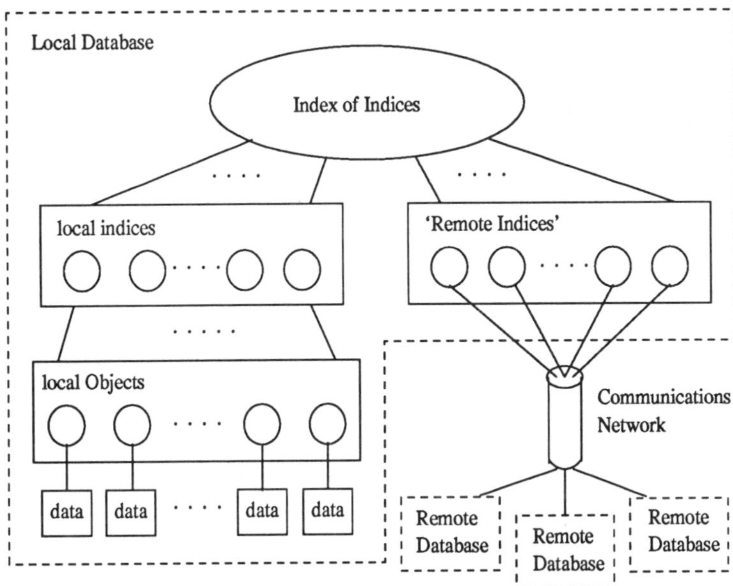

Figure 13.4: Architecture of the model database

of objects and all indices without user intervention, so that when an object is locally created, modified or destroyed the changes are automatically propagated to all database processes which have expressed an interest in objects of that class, and are reflected in all indices, local or remote, which contain or should contain the object. Thus, each client process can behave as if its own copy of an object was the only one and use a variety of indexing systems without explicitly having to maintain them. This process, illustrated in Fig. 13.4, is achieved as follows:

All indices present a uniform interface to the user with primitives such as object insertion and deletion and a generalised search which performs a user-defined action on every object in the index which fits a user-supplied 'Object Template'. Object Templates are data structures designed to act as 'Object filters' and are used uniformly throughout the system to select interesting classes of objects. In particular, every index has a private Template which determines which objects it will accept. Moreover, all indices within the system are actually Objects in their own right and may themselves be stored in system indices. This allows local indices to be maintained by keeping an index of indices and searching it for the indices which should be updated when a particular object is altered. Similarly, a remote process with an interest in a class of objects lodges a 'Remote Index' in the local index of indices; this acts as a gateway to the remote process, transmitting appropriate classes of changes across a communications network to its parent. In both cases Object Templates are used for object selection and index search pruning.

We have not yet addressed the difficult problems of inconsistency or deadlock which arise when several processes simultaneously attempt to modify an object, however these are not expected to be severe in our domain as the few parts of the database which are frequently changed (such as robot states and positions) tend to be under the control of a single process.

13.2.2 Implementation

At present the system is configured as a central database serving a number of sensor, controller and planning processes, each with embedded local database code. It is written in C and runs on (a network of) SUN workstations under UNIX, using SUN's Remote Procedure Call and External Data Representation mechanisms for network (INTERNET) communications in a machine-independent data format. The present implementation is only a prototype, but it is designed to make the following changes straightforward to implement: switching to an object-oriented language such as C++ would simplify the code considerably and allow a richer object hierarchy; UNIX responds too slowly for realistic robot control and parts of the system will have to be ported to a different operating system (e.g., to run directly under the architecture discussed in Chapter 3); and the Remote Procedure Call mechanism is not well adapted to our purposes and should be replaced with a lower level network protocol in a future version.

All of the basic types of indices are implemented using binary trees. The name and serial number indices are configured forms of a general purpose binary-tree index (with additional tree-balancing code) and the two and three dimensional spatial indices have a layered binary tree structure as follows:

Recall that spatial indices, being Objects, are provided with spatial configurations and bounding boxes; all indexing calculations are performed in index-configuration-aligned coordinates so that spatial indices with arbitrary spatial alignment are possible. For an object to be stored, its bounding box (in aligned coordinates) must overlap the aligned bounding box of the spatial index. If this is so, the index box is repeatedly bisected in the x-direction, the half containing the object being retained, until the object box is cut by a bisection rectangle; this rectangle is then bisected in the y-direction until the object box is cut by a bisection line; and the line is bisected in the z-direction until a bisection point falls within the object box; the object is then stored in a list of the objects associated with this bisection point. Thus, a three dimensional spatial index is a binary (x–) tree of binary (y–) trees of binary (z–) trees of lists of objects, as illustrated in Fig. 13.5. Similarly, a two dimensional index is an x-tree of y-trees of object lists.

As is often the case with bisection-based spatial indices, each tree node is associated with a unique bounding box (the box which is bisected at the node), a unique spatial point (the centroid of the node box), and a unique set of three binary fractions (the coordinates of the node centre in the system normalising the index box to $[0,1] \times [0,1] \times [0,1]$), which can conveniently be used for node identification and tree traversal. Objects are stored at the node corresponding to the smallest node box containing the object and the first node centre to fall inside the object box.

Fig. 13.6 illustrates a two dimensional spatial index in operation. The polygons and bounding boxes of the objects are shown, and the dotted lines are node bisectors; thus, 'pallet1' is cut by the third layer of bisection on the x-axis (which also cuts 'pallet2') and the third layer bisection on the y-axis, and its fractional node coordinates are $(.001, .011)$.

To prevent very small or thin objects falling too deep in the tree a resolution cut off is usually applied, objects being stored in the smallest remaining node box containing them. It is possible to alter the order of the node bisections without affecting the depth of any object in the tree.

Note that the node boxes repeatedly overlap one another: this means that finding the objects overlapping a given box requires a search of all of the parents (containing nodes) and many of the children (contained nodes) of the box node; however the search is easily implemented by recursively eliminating nodes (and hence subtrees) which do *not* overlap the given box and this requires only a single coordinate comparison at each node.

There are of course many alternative ways of organising such a bisection-based spatial index; in particular quad-tree or oct-tree based representations [179] (in which the node box is divided into quadrants or octants at each level) might be thought to be the obvious choice. The chosen bin-tree system has a number of advantages over

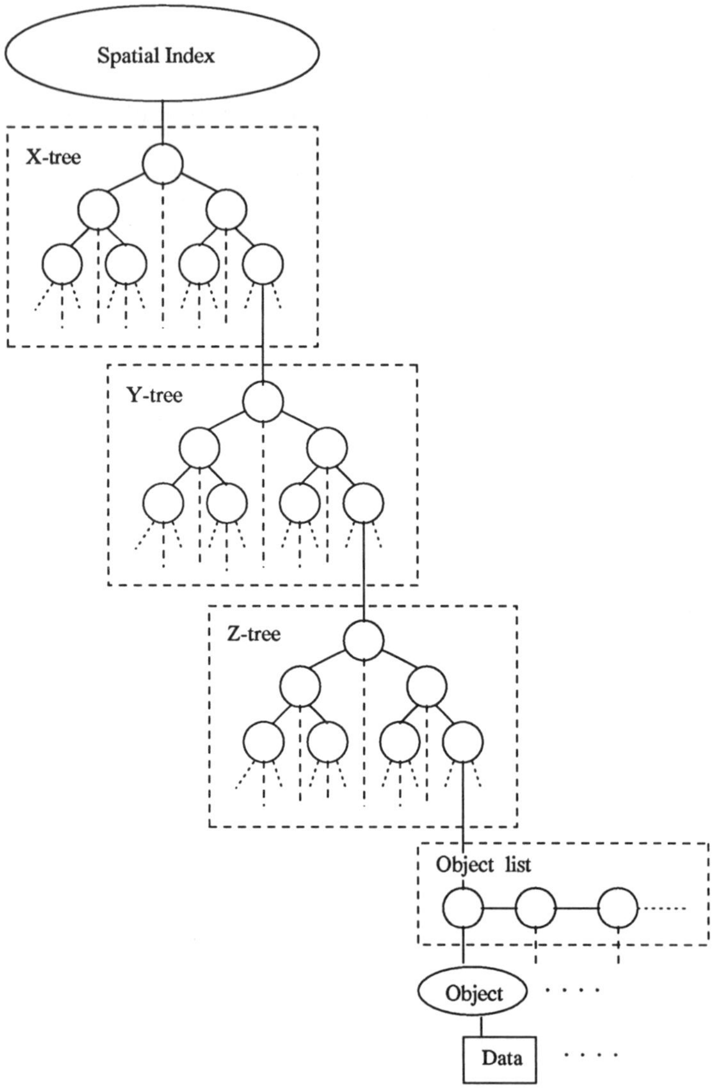

Figure 13.5: Spatial indices have a layered binary tree structure

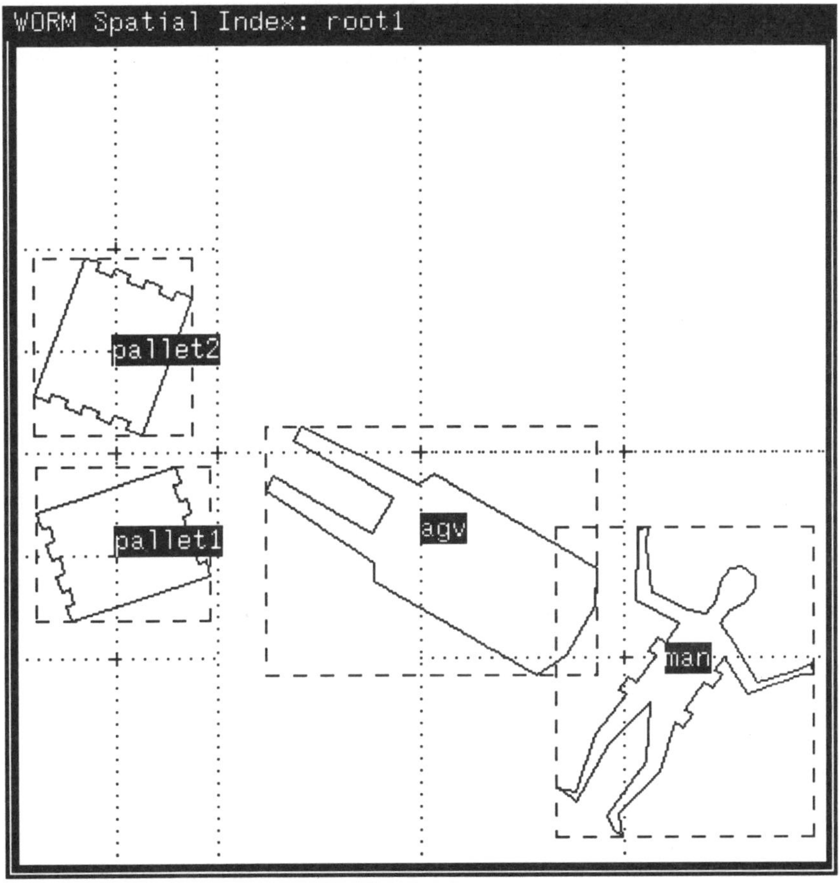

Figure 13.6: A two dimensional spatial index in action

these:

1. The bin-tree representation has node-boxes of arbitrarily high aspect ratio and is therefore extremely efficient at circumscribing long thin objects such as walls, corridors and rows of storage racks, provided they are aligned with the index axes. Such objects are obviously very common in factory-like environments. By comparison, the quad– and oct-tree structures (at least in their simplest forms) force all of the tree nodes to have the same aspect ratio as the original index-box and may therefore be rather poor at object localisation, particularly if the region being indexed is far from square.

2. The bin-tree representation leads to slightly simpler and more compact code and is trivial to adapt to indices of any dimensionality; in fact the two and three dimensional indices share the same code in our system.

3. Although it is true that quad– or oct-trees tend to be shallower than their binary equivalents, it does not follow that they are much faster to traverse or more compact. Our experience is that access time depends more on the total amount of bisection required than on the tree depth because most of the time is spent doing bisection arithmetic, and that (at least in the robot modelling domain) the higher branching ratio trees are actually less compact because most of the node pointers are never used. One caveat is that because the bin-tree system uses a larger number of smaller nodes it is comparatively more dependent on the efficiency of the memory allocator for nodes.

In practice this layered binary tree system has proved to be highly efficient at object localisation: simple, fast and reasonably compact, despite the fact that quite deep trees are generated.

13.2.3 Architecture—Summary

Many robots would be significantly more flexible, powerful and easy to use if they were fitted with a geometric model of their surroundings, and this is particularly true of autonomous robot vehicles. When such a world model is supported by a properly designed database the further advantages of a modular, data-driven, database-centred system architecture with a clean, consistent user interface are available, and if this database is also able to support distributed processes the resulting system forms the basis of a powerful integrated multi-robot environment. A particular advantage of this architecture is the ease with which it can be extended as new expertise becomes available.

Running such a model-based, data-driven system makes considerable demands of the underlying database, however we feel that these will largely be met by the modular object-oriented robot database system we envisage.

13.3 Routemap Extraction in an Environment having Manthattan Geometry

Both the off-line path planner in Chapter 11 and the on-line planner in Chapter 4 can use data about the natural corridors in an environment. In many applications these could sensibly be entered by hand, but we have developed an automated process generating such corridors, that can either be used instead of or to assist a human operator.

13.3.1 Extracting Corridors

The input data for the routemap extraction algorithm takes the form of a 2-dimensional polygonal boundary model of the environment, which may be manually derived, or obtained by projection from a 3-dimensional factory model. The model must have a Manhattan geometry, i.e., all boundaries must lie parallel to one of the coordinate axes. (A variation on this algorithm that efficiently handles non-Manhattan environments is detailed in [82].)

The first action of the algorithm is to break up this model into a set of boundaries, each with an associated outward pointing normal vector. This set may then divided into two subsets, those boundaries parallel to the x-axis, and those parallel to the y-axis. Let these sets be denoted X and Y respectively. These two subsets are then sorted, on the y value for set X, and on the x value for Y.

Consider first the set X of obstacle boundaries that are parallel to the x-coordinate axis. We now proceed using a sweep line algorithm [166] as follows. Consider an imaginary line, infinite in extent and parallel to the x-axis, moving from $y = -\infty$ to $y = +\infty$. For any value of y, there exists a set of non-overlapping segments of this line which lie within free space. Initially, at $y = -\infty$, this is the empty set (we assume that free space is bounded, and hence there is no free space outside the boundary). As the imaginary line is swept in the direction of increasing y, the previously sorted set of obstacle boundaries are encountered in turn. When a boundary is encountered which has a normal vector in the positive y direction, a new free space segment is created. If this new segment is contiguous with any existing segments, they are merged to form a single new segment. When a boundary is encountered which has a normal vector in the negative y direction, the remainder of the segment after the range of the obstacle boundary is removed forms none, one or two new segments.

By performing these operations for all boundaries in the environment parallel to the x-axis, free space may be partitioned into a set of rectangular regions swept out by each of the segments described above, as shown in Fig. 13.7(a). The boundaries of these regions are an extension of the corresponding member(s) of X. The regions may be of two types, depending upon the direction of the normal vectors associated with the obstacle boundaries from which the region's boundaries are derived. If both these normals are directed *into* the region, it is labelled as a *corridor*, and the midline of

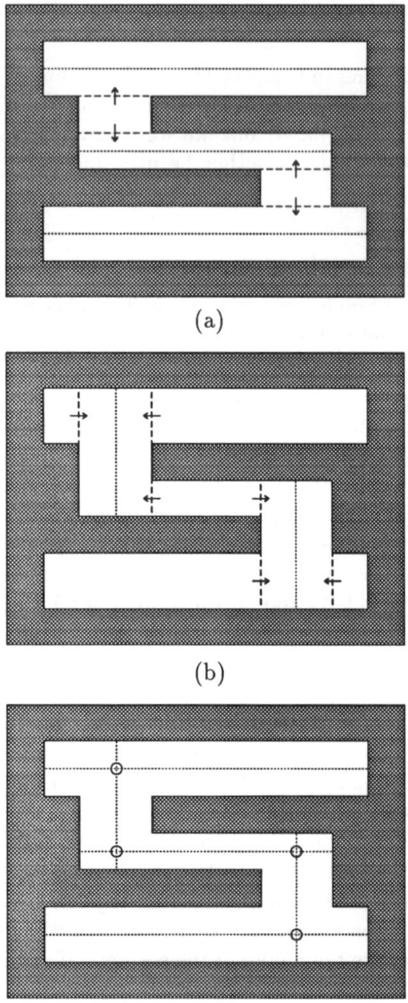

(a) (b) (c)

Figure 13.7: Illustration of the regions and corridors extracted by the algorithm: (a) parallel to x-axis, (b) parallel to y axis. The intersection of these corridors is seen in (c). (Dotted lines represent the centerline of a corridor, dashed lines region boundaries)

the corridor (depicted as a dotted line in Fig. 13.7) is computed. These midlines will later be used to build the graph structure used for global planning. If either (or both) of the normals are directed *outward* from the region, then a pointer is associated with that region. This pointer indicates the adjacent region, as indicated by the arrows in Fig. 13.7. The list of regions generated, and the pointers associated with them, may be used to compute, for any point in free space, a path onto one or more of the corridor midlines which does not intersect any obstacles.

This process may now be repeated for set Y (e.g., Fig. 13.7(b)), providing a further set of regions, and completing the set of corridors.

13.3.2 Structure of the Corridor Graph

A searchable routemap may now be built by computing the intersections of the corridor midlines derived above, see Fig. 13.7(c). The graph structure used to represent the routemap is based upon two primitives:

1. *corridors*, which have two components,
   (a) The midline, described by its end points, and
   (b) A list of the nodes which lie on this corridor
2. *nodes*, which also have two components,
   (a) The location of the node, and
   (b) The two corridors which intersect at this node.

Note that the corridors are not split into smaller arcs connecting pairs of nodes, but are represented as a single unit. Fig. 13.8 shows a pair of example routemaps.

13.3.3 Computational Complexity

The sorting process required is accomplished with $O(n \log n)$ worst case time complexity, where n is the number of obstacle boundaries. A modified merge sort is used, which may exploit existing ordered sequences in the input. This is likely to be common in sparse environments having small obstacles, since the boundaries from a single obstacle will occur together in the input.

The subsequent processing of the boundaries to extract corridors may also be performed in time $O(n \log n)$, if the set of free space segments is stored in a balanced tree structure (such as AVL or 2-3 trees) for $O(\log n)$ access time.

Determining the connectivity of the corridors so generated in order to build a searchable roadmap requires time $O(k + n)$ (where $k < n^2$ is the size of the output), although the intersection tests required are relatively fast. This gives the overall worst case complexity of $O(k + n \log n)$. For environments which do not have a Manhattan geometry, a second approach is described in [82]. The principal of this second algorithm is to decompose free space into a set of non-overlapping convex fragments, whilst attempting to minimise the *number* of fragments generated.

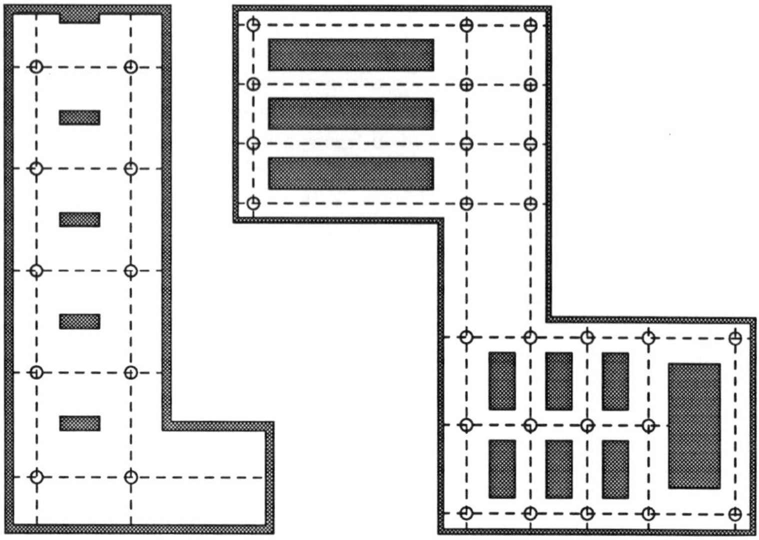

Figure 13.8: Example routemaps

13.3.4 Searching the Routemap

Given arbitrary start and goal locations, the global planner must first find collision free paths connecting these points to one or more of the corridor midlines. This can be readily achieved using the two sets of regions derived earlier.

Each of the two terminal points must fall within one region in each of the two region sets. These regions are either a corridor, or have a pointer to one or more adjacent regions. By following the region pointers, a number of corridors will be found. By virtue of the manner in which the region pointers were assigned, the corridors found have the property that a line perpendicular to the corridor midline and passing through the terminal point will connect that point and the corridor with a path which does not intersect obstacles.

Nodes corresponding to the two terminal points, and the locations at which the perpendiculars described above intersect the corridors of the precalculated graph, are thus added to the graph along with the appropriate connecting edges.

All that remains to find a path between the two points is to search the graph, for which we use an A* search [89]. The weight associated with a path is calculated as the total distance travelled, *plus* a constant cost for each node in the graph visited. This makes use of an unusual feature of the graph in that each corridor is represented as one unit having several nodes along it, rather than as a set of shorter edges connected at the nodes. It is hence not necessary for every node on the corridor to be visited

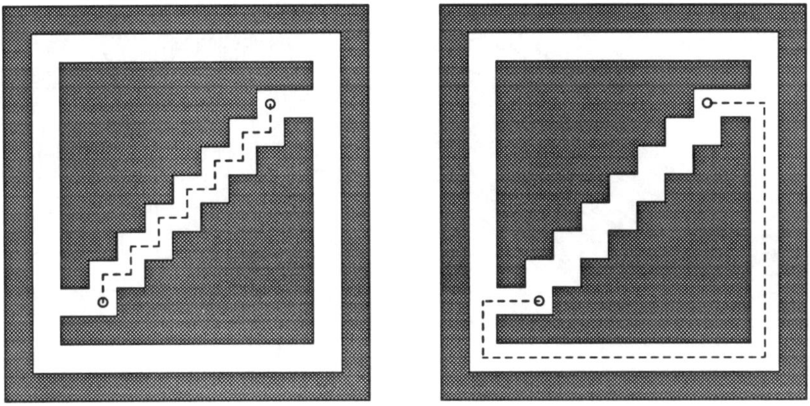

(a) (b)

Figure 13.9: The effect of node penalty upon path choice; in (a) the node penalty is small, and the shorter path is chosen. In (b), it is large, so the path with fewer manœuvres is prefered

when traversing the corridor. Indeed, for the lowest cost path found by the search the only nodes visited will be the start and goal nodes, plus those nodes at which a turn must be made. The constant node cost is, then, used to represent the cost of performing such a manœuvre. Fig. 13.9 illustrates the effect of different node costs; in Fig. 13.9(a) the node cost is small, and hence the shortest path between the obstacles is chosen. In Fig. 13.9(b), the node cost is set to a higher penalty, causing the path with less manœuvering to be selected in preference.

Since the graph has a Manhattan geometry, the heuristic used by the A* search is the distance between the current node location and the goal under the L_1 metric[1].

It is worth noting at this point that the algorithm described above may easily be modified to allow multiple lanes of traffic, where space permits. If the required width of the lanes is specified, instead of simply adding the midline of a corridor to the graph, the width of the corridor region can be used to evaluate the number of parallel lanes which may be fitted in the available space.

13.4 A Large Example

In order to illustrate that our software can successfully deal with large models, we visited Hornby plc and obtained from them a plan of their factory. Hornby are a large

[1]The L_1 distance between points (x_1, y_1) and (x_2, y_2) is $|x_1 - x_2| + |y_1 - y_2|$.

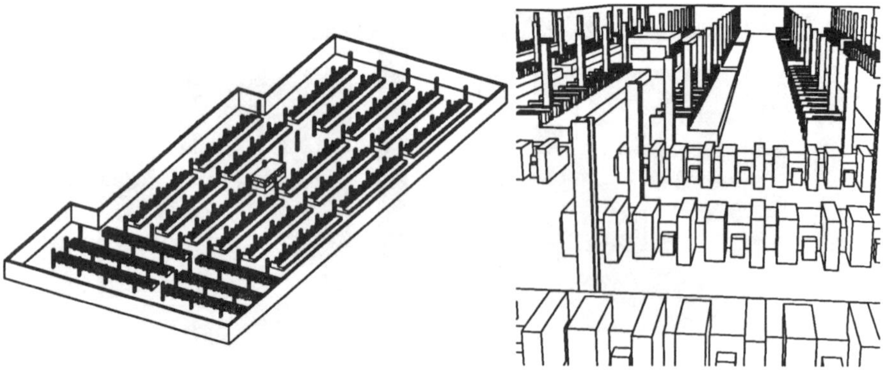

(a) (b)

Figure 13.10: Views of the Hornby factory, modelled with ROBMOD

toy manufacturer based in Westgate, and their main factory area is an old building which is mainly given over to assembly by humans. As we were more concerned with generating a model that was realistic in terms of its size and its variation in geometric detail, rather than modelling the individual workstations, we used a single generic model of the assembly lines, which gives the resultant model a sense of order not to be found in the real factory! In fact, we initially built the model as a three-dimensional model using ROBMOD (Fig. 13.10(a)), which allowed us to generate synthetic views (Fig. 13.10(b)).

A two-dimensional floorplan model was then generated directly from this model using a *projection* function that we added to ROBMOD, generating the world model shown in Fig. 13.11(a). As can be seen the geometric detail has not been lost, with a 500-to-1 difference in the dimensions of the long wall compared to the indentations in the I-beams! This model size is around ten thousand edges, the binary division strategy employed in the world model sucessfully coped with this structure. Also, the corridor extraction routine can also be run on this example. A parameter in the extractor filters out the very small corridors that could be derived into each individual seating station on the assembly lines, and setting the parameter to a value suitable for the vehicle produces the (quite satisfactory) set of corridors shown in Fig. 13.11(b). Of course, this set is unlikely to be accepted in a workspace as is, as it takes no account of the problems of segregating people and robots, but it does form a sensible starting point for a human editor.

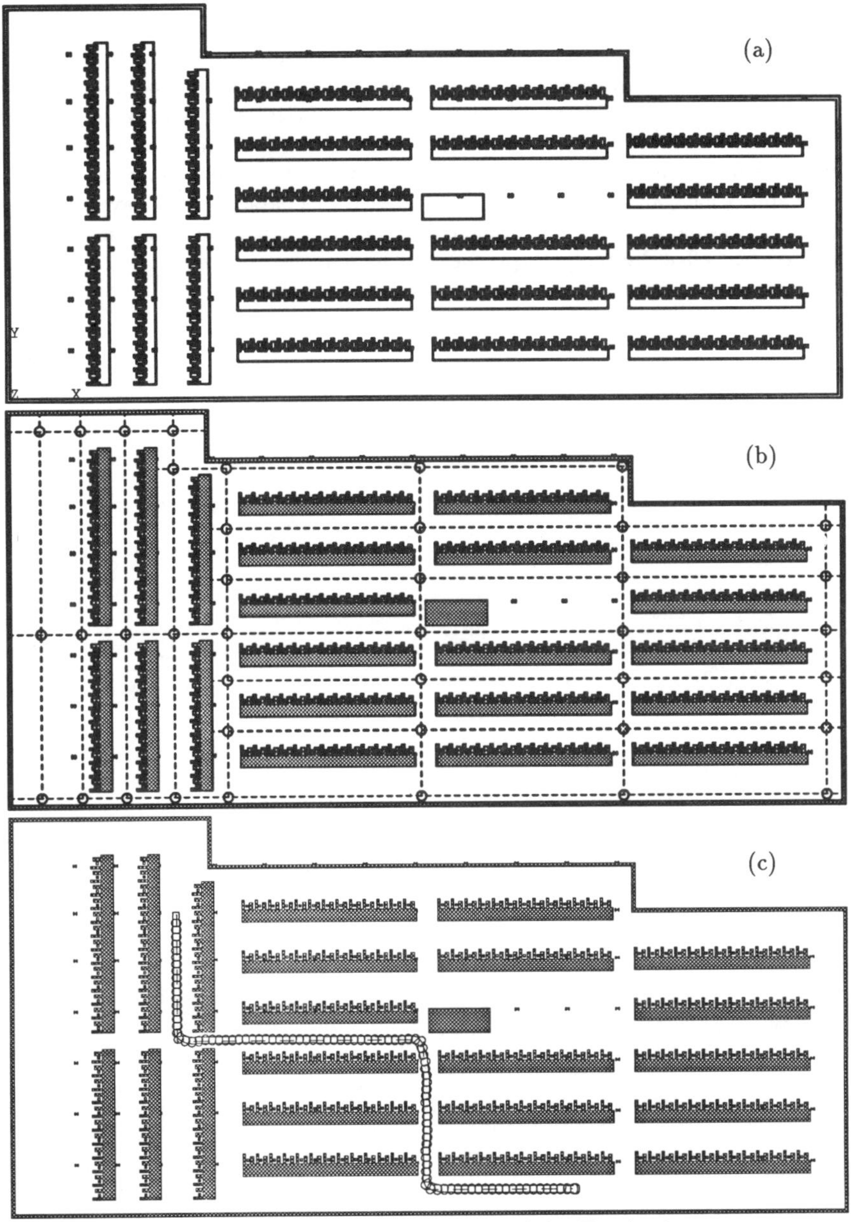

Figure 13.11: Two-dimensional plans of the Hornby factory

13.5 Summary

We have shown that it is possible to develop a distributed geometric database that is optimised for use with robot vehicles in an industrial setting. The database itself is based on a binary division strategy, which is efficient and is capable of storing a wide variety of data that is *spatially localised* (i.e., data that can be assigned a definite—though not necessarily precise—location). The various planners of the last few chapters have all been designed to work with this database. We have also shown how it is possible to automatically extract 'corridors' from the database, which are used to great effect by the path planner of Chapter 11.

Epilogue

The work in this book does not describe a single project but rather a common integrating theme that has run throughout a large part of the work in the Robotics Research Group. In compiling the book, the editors have taken a snapshot at around the end of 1992 of three major strands of work on the Oxford AGV. We have benefited through gaining a better understanding of the relationship between different areas of the project and of the scope they offer. We hope that the book will prove useful, both as a record of work performed and as a cookbook for people interested in developing mobile robot systems.

This book has, of necessity, only touched on some of the the work carried out more generally in Oxford which is relevant to autonomous vehicles. The sheer number of people involved has lead to a move towards more specialised software and hardware platforms. One key strand is the development of a family of modular vehicles, for which the sensor described in Chapter 7 was developed, which as well as acting as general platforms are becoming useful research tools for investigation into the control of redundant mechanisms. The latest vehicle controllers are becoming increasingly sophisticated with the incorporation of the decision theoretic ideas of chapter 4, and of sensor integration and management modules. Sensor management frameworks have been developed for the LICA based on various levels of distribution of function, and we soon expect to demonstrate sensor management in reactive planning using a number of the sensors described in this book. The work on model based planning was the starting point for projects on planning for robot manipulators, and (with the problem of a single vehicle essentially solved) on the co-ordination and scheduling of a fleet of vehicles. In addition over the last few months there have been exciting advances in the application of an active stereo head to real time control.

Thus the original work has spawned a wide variety of further topics, now being actively pursued in Oxford and elsewhere, and fully justified the original intention of using the vehicle to bring together work in different fields. We hope that we have conveyed some of the excitement of the challenges in this research area.

SAC & PJP
December 1993

Bibliography

[1] M. Adams, H. Hu, and P.J. Probert. Towards a real-time architecture for obstacle avoidance and path planning in mobile robots. In *Proc. IEEE Int. Conf. Robotics and Automation*, pages 584–589, Cincinatti, USA, 1990.

[2] M.D. Adams. *Optical Range Data Analysis for stable target pursuit in mobile robotics*. PhD thesis, Oxford University, 1992.

[3] M.D. Adams and P.J. Probert. Logical sensor development for planning and navigation for a mobile robot. submitted to International Journal of Robotics Research, 1993.

[4] J.R. Albus, A.J. Barbera, and R.N. Nagel. Theory and practice of hierarchical control. In *Proc. of the 23th IEEE Computer Society Int. Conf.*, pages 18–27, New York, September 1981.

[5] J.R. Albus, C.R. Mclean, C.R. Barbera, and M.L. Fitzgerald. An architecture for real-time sensory-interactive control of robots in a manufacturing facility. In *Proc. of the 4th IFAC/IFIP Symposium in Information Control Problems in Manufacturing Technology*, pages 81–90, Gaithersburg, 26-28 October 1982.

[6] J.R. Allen and K. Kennedy. PFC: A program to convert Fortran to parallel form. In *Proc. of the IBM Conference on Parallel Computers and Scientific Computations*, March 1982.

[7] R.L. Andersson. Living in a dynamic world. In *Proc. of the ACM-IEEE Fall Joint Computer Conference*, pages 97–104, 1986.

[8] R.L. Andersson. *A Robot Ping-Pong Player*. MIT Press, 1988.

[9] R.C. Arkin. Motor schema-based mobile robot navigation. *Int. J. Robotics Research*, 8(4):92–112, 1989.

[10] E. Badreddin. A recursive control structure for mobile robots. In G. Schmidt, editor, *Information Processing in Autonomous Mobile Robots*, pages 171–185. Springer-Verlag, 1991.

[11] J. Barraquand and J. C. Latombe. Robot motion planning: A distributed representation approach. Technical Report STAN-CS-89-1257, Department of Computer Science, Stanford University, 1989.

[12] J. Barraquand and J. C. Latombe. Robot motion planning using a distributed representation approach. *Int. J. of Robotics Research*, 10(6), December 1991.

[13] B. Barshan and R. Kuc. Diferentiating sonar reflections from corners and planes by employing and intelligent sensor. *IEEE Transactions on Pattern Analysis and Machine Intelligence*, 12(6):560–569, 1990.

[14] B. Barshan and R. Kuc. ROBAT: A sonar based mobile robot for bat-like prey capture. *Proceedings of IEEE Conference of Robotics and Automation*, pages 274–279, 1992.

[15] J. Berger. *Statistical Decision Theory and Bayesian Analysis.* Springer Verlag, 1985.

[16] P. J. Besl and R. C. Jain. Surface characterization for three-dimensional object recognition in depth maps. Technical Report RSD-TR-20-84, Centre for Research on Integrated Manufacturing, University of Michigan, 1984.

[17] S.S. Blackman. *Multiple Target Tracking with Applications to Radar.* Artec House, 1986.

[18] A. Blake, D. McCowen, H.R. Lo, and D. Konash. Epipolar geometry for trinocular active range sensors. In *Proc. British Machine Vision Conf.*, 1990.

[19] A. Blake and A. Zisserman. *Visual Reconstruction.* MIT Press, 1987.

[20] R. J. Blissett. Retriving 3D information from video for robot control and surveillance. *Electronics & Communication Engineering Journal*, pages 155–163, August 1990.

[21] H. Blum. Biological shape and visual science. *Journal of Theoretical Biology*, 38:205–287, 1973.

[22] H. Blum and R. N. Nagel. Shape description using weighted symmetric axis features. *Pattern Recognition*, 10:167–180, 1978.

[23] R. C. Bolles and P. Horaud. 3DPO: A three-dimensional part orientation system. *International Journal of Robotics Research*, 5(3):3–26, 1986.

[24] J. Borenstein and Y. Koren. Obstacle avoidance with ultrasonic sensors. *IEEE J. Robotics and Automation*, 4(2):213–218, 1988.

[25] J. Borenstein and Y. Koren. Real time obstacle avoidance for fast mobile robots. *IEEE Transaction on Systems, Man and Cybernetics*, 19(5):1179–1187, 1989.

[26] J. Borenstein and Y. Koren. Analysis of control methods for mobile robot obstacle avoidance. In *IEEE International Workshop on Intelligent Motion Control*, pages 457–462, August 1990.

[27] J. Borenstein and Y. Koren. Real time obstacle avoidance for fast mobile robots in cluttered environments. In *Proc. IEEE International Conference on Robotics and Automation*, pages 572–577, 1990.

[28] J. Borenstein and Y. Koren. Noise rejection for ultrasonic sensors in mobile robot applications. In *RSJ/IEEE International Workshop on Intelligent Robots and Systems*, pages 1727–1732, 1992.

[29] J. M. Brady, J. Ponce, A. Yuille, and H. Asada. Describing surfaces. *Computer Vision, Graphics, and Image Processing*, 32:1–28, 1985.

[30] J M Brady and H Wang. Vision for mobile robots. *Phil. Trans. R. Soc. Lond. B*, 337:341–350, Sept. 1992.

[31] Michael Brady, Stephen Cameron, Hugh Durrant-Whyte, Margaret Fleck, David Forsyth, Alison Noble, and Ian Page. Progress towards a system that can acquire pallets and clean warehouses. In *Fourth Int. Symp. Robotics Research*, pages 359–374, Santa Cruz, August 1987.

[32] R. A. Brooks. Solving the find-path problem by good representation of free space. *IEEE Trans. Systems, Man and Cybernetics*, 13:190–197, 1983.

[33] R. A. Brooks. A robust layered control system for a mobile robot. *IEEE Trans. Robotics and Automation*, 2(1):14–23, 1986.

[34] R.A. Brooks. A layered intelligent control system for a mobile robot. In *Third Intl. Symp. Robotics Research*, pages 365–372, 1986.

[35] R.A. Brooks. Self calibration of motion and stereo vision for mobile robots. In *Fourth Intl. Symp. Robotics Research*, 1987.

[36] Rodney A. Brooks. Autonomous mobile robots. In W.E.L. Grimson and R.S. Patil, editors, *AI in the 1980s and Beyond.* The M.I.T Press, 1987.

[37] M.K. Brown. Feature extraction techniques for recognising solid objects with an ultrasonic range sensor. *IEEE J. Robotics and Automation*, 1(4):191–205, 1985.

[38] M. Brownlow. *An Optical Time of Flight Rangefinder for Mobile Robot Navigation.* PhD thesis, Oxford University, 1993.

[39] C. E. Buckley. A foundation for the "flexible-trajectory" approach to numeric path planning. *Int. J. Robotics Res.*, 8(3):44–64, June 1989.

[40] C. E. Buckley and L. J. Leifer. A proximity metric for continuum path planning. In *Int. Joint Conf. Artificial Intelligence*, pages 1096–1102, Los Angeles, August 1985.

[41] Alan Burns. *Programming in OCCAM 2.* Addison-Wesley Publishing Company, 1988.

[42] A. Cameron. *A Bayesian Approach to Optimal Sensor Placement.* PhD thesis, University of Oxford, U.K., 1989.

[43] S. A. Cameron. A study of the clash detection problem in robotics. In *IEEE Int. Conf. Robotics and Automation*, 1985.

[44] S. A. Cameron. A geometric database for the oxford autonomous guided vehicle. In B. Ravani, editor, *CAD Based Programming for Sensory Robots.* Springer-Verlag, 1988.

[45] S. A. Cameron and J. C. Aylett. ROBMOD users guide. Software report, Department of Artificial Intelligence, University of Edinburgh (U.K.), 1987.

[46] S. A. Cameron and R. K. Culley. Determining the minimum translational distance between two convex polyhedra. In *Int. Conf. Robotics & Automation*, pages 591–596, San Francisco, April 1986.

[47] Stephen Cameron and Jon Aylett. ROBMOD: A geometry engine for robotics. In *Int. Conf. Robotics & Automation*, pages 880–885, Philadelphia, April 1988.

[48] J. Canny. Collision detection for moving polyhedra. *IEEE Trans. Pattern Analysis and Machine Intelligence*, PAMI-8:200–209, 1986.

[49] D Charnley and R Blisset. Surface reconstruction from outdoor image sequences. *Image and Vision Computing*, 7(1):10–16, 1989.

[50] R. Chatila and J.P. Laumond. Position referencing and consistent world modeling for mobile robots. In *Proc. IEEE Int. Conf. Robotics and Automation*, pages 138–143, 1985.

[51] W. F. Clocksin and C. S. Mellish. *Programming in Prolog.* Springer-Verlag, Berlin, 1981.

[52] R.F. Cmelik, N.H. Gehani, M. Plotnick, and W.D. Roome. Concurrent C project. Technical report, A collection of reports on concurrent C, 1986.

[53] J. Connell. *Minimalist Mobile Robotics*, volume 5 of *Perspectives in Artificial Intelligence*. Academic press, INC, 1990.

[54] I.J. Cox. Blanche: an autonomous robot vehicle for structured environments. In *IEEE Conf. on Robotics and Automation*, pages 978–982, 1988.

[55] J.L. Crowley. Navigation for an intelligent mobile robot. *Proc. IEEE Int. Conf. Robotics and Automation*, 1:31–41, 1985.

[56] J.L. Crowley. World modelling and position estimation for a mobile robot using ultrasonic ranging. In *IEEE Conf. on Robotics and Automation*, pages 674–680, 1989.

[57] M. Daily et al. Autonomous cross-country navigation with the ALV. In *Proc. IEEE Int. Conf. Robotics and Automation*, pages 718–726, 1988.

[58] W. Dickson. Feature grouping in a hierarchical probabilistic network. In *Proc. British Machine Vision Conf.*, pages 13–18, 1990.

[59] M. Donath and T. L. Anderson. Animal behaviour as a paradigm for developing robot autonomy. *Robotics and autonomous systems*, 6:145–168, 1990.

[60] A. Elfes. Sonar-based real world mapping and navigation. *IEEE J. Robotics and Automation*, RA-3(3):249–265, 1987.

[61] A. Elfes. Integration of sonar and stereo range data using a grid-based representation. In *Proc. IEEE Int. Conf. Robotics and Automation*, pages 727–732, 1988.

[62] A. Elfes and S.N. Talukdar. A distributed control system for the CMU rover. In *Int. Joint Conf. Artificial Intelligence*, page 830, 1983.

[63] H. Everett and A. Flynn. A programmable near-infrared proximity detector for mobile robot navigation. In *Proc. SPIE Conference on Advances in Intelligent Robotics Systems*, volume 727, pages 221–230, 1986.

[64] H.R. Everett. Survey of collision avoidance and ranging sensors for mobile robots. *Robotics and Autonomous Systems*, 5:5–67, 1989.

[65] O. D. Faugeras and M. Hebert. The representation, recognition, and locating of 3D objects. *International Journal of Robotics Research*, 5(3):27–52, 1986.

[66] R. B. Fisher. *From surfaces to objects: computer vision and three dimensional scene analysis.* John Wiley and Sons, 1989.

[67] R. B. Fisher and M. J. L. Orr. Geometric reasoning in a parallel network. *International Journal of Robotics Research*, 10(2):103–122, 1991.

[68] L. De Floriani. Surface representation based on triangular grids. *The Visual Computer*, 3:27–50, 1987.

[69] A.M. Flynn. Combining sonar and infrared sensors for mobile robot navigation. *International Journal of Robotics Research*, 7(6):5–14, 1988.

[70] M.J. Flynn. Some computer organizations and their effectiveness. *IEEE Trans. on Computers*, C-21(9):948–960, 1972.

[71] B. R. Fox and F. G. Kempf. Opportunistic scheduling for robotic assembly. In E. L. Fisher and O. Z. Maimon, editors, *Industrial Engineering, Seleted Readings.* MIT Press, 1986.

[72] G C Fox, M A Johnson, G A Lyzenga, S W Otto, J K Salmon, and D W Walker. *Solving problems on concurrent Processors, Volume 1.* Prentice-Hall, Englewood Cliffs, 1988.

[73] T. Fraichard. Smooth trajectory planning for a car in a structured world. In *IEEE Int. Conf. Robotics and Automation*, pages 318–323, Sacramento, April 1991.

[74] GEC-FAST. LCS2 road system user guide. Technical Report T299 Issue3 12.85, GEC Industrial Controls Ltd., Rugby (UK), 1986.

[75] GEC-FAST. The Oxford University AGV. Technical report, GEC Electrical Projects Ltd, Factory Automation Division, Rugby (UK), 1987.

[76] GEC-FAST. The GEC AGV control system. Technical Report MTR Issue B, GEC Electrical Projects Ltd, Factory Automation Division, Rugby (UK), 1988.

[77] G. Giralt. Research trends in decisional and multisensory aspects of third generation robots. In *2nd International Symposium on Robotics Research*, 1984.

[78] G. Goto and A. Stentz. The CMU system for mobile robot navigation. In *IEEE Conference on Robotics and Automation*, pages 99–105, 1987.

[79] V. Graefe. Dynamic vision systems for autonomous mobile robots. In *RSJ/IEEE International Workshop on Intelligent Robots and Systems*, pages 12–23, 1989.

[80] V. Graefe and U. Jacobs. Detection of passing vehicles by a robot car driver. In *RSJ/IEEE International Workshop on Intelligent Robots and Systems*, pages 391–396, 1991.

[81] W. E. L. Grimson. *Object recognition by computer: the role of geometric constraints.* MIT Press, 1991.

[82] Tony Hague. *Motion Planning for Autonomous Guided Vehicles.* PhD thesis, Oxford University, 1993.

[83] P.B. Hansen. The programming language Concurrent Pascal. *IEEE Transactions on Software Engineering*, SE-1(2):199–206, June 1975.

[84] R. M. Haralick. Edge and region analysis for digital image data. *Computer Vision, Graphics, and Image Processing*, 12:60–73, 1980.

[85] S.Y. Harmon. The ground surveillance robot(GSR): An autonomous vehicle designed to transit unknown terrain. *IEEE J. Robotics and Automation*, 3:266–279, 1987.

[86] S.Y. Harmon. Architectures for real-time intelligent control systems: Concepts, approaches, methodologies. In *Record of the Working Group at the Workshop on Architectures for Intelligent Control Systems*, Marriott Key Bridge Hotel, Arlington, VA, August 1991.

[87] C Harris. Determination of ego-motion from matched points. In *Proc. 3rd Alvey Vision Conference*, pages 189–192, Cambridge, Sept. 1987.

[88] C G Harris and J M Pike. 3D positional integration from image sequences. In *Proc. 3rd Alvey Vision Conference*, Cambridge, Sept. 1987.

[89] P. Hart, N. J. Nilsson, and B. Raphael. A formal basis for the determination of minimum cost paths. *IEEE Trans. Systems, Man and Cybernetics*, 4(2):100–107, July 1968.

[90] M. Hebert and E. Krotkov. 3-D measurements from imaging laser radars: how good are they? In *RSJ/IEEE International Workshop on Intelligent Robots and Systems*, pages 359–364, 1991.

[91] M. Hebert, E. Krtokov, and T. Kanade. A perception system for a planetary explorer. In *Proc. of 28th IEEE Conference on Decision and Control*, pages 1151–1156, 1989.

[92] M. Herman and J.R. Albus. Overview of the multiple autonomous underwater vehicles (MAUV) project. In *Proc. IEEE Int. Conf. Robotics and Automation*, pages 618–620, Philadelphia, 24-29 April 1988.

[93] C.A.R. Hoare. Communicating sequential processes. *Communications of the ACM*, 21(8):666–677, 1978.

[94] C.A.R. Hoare. *Communicating Sequential Processes.* Prentice Hall, 1985.

[95] H. Hofler, G. Schmidkte, J. Molnar, and L. Unger. 3D measurement techniques for robotics and quality control with structured illumination. In *25th ISATA Proceedings*, pages 81–88, 1992.

[96] H. Hu. *Dynamic Planning and Real-time Control for a Mobile Robot.* PhD thesis, University of Oxford, 1992.

[97] H. Hu, M. Brady, and P. Probert. Coping with uncertainty in control and planning for a mobile robot. In *IEEE/RSJ Int. Workshop on Intelligent Robots and Systems*, Osaka, Japan, 1991.

[98] H. Hu, M. Brady, and P.J. Probert. Navigation and control of a mobile robot amongst moving obstacles. In *30th IEEE Conference on Control and Decision*, pages 698–703, 1991.

[99] H. Hu, P. Probert, and M. Brady. Transputer architecture for real-time control of a mobile robot. In *Int. Conf. on Applications of Transputers 91*, pages 406–411, Glasgow, U.K., 1991.

[100] H. Hu and P.J. Probert. Distributed architectures for sensing and control in obstacle avoidance for autonomous vehicles. In *IARP Int. Conf. Multi-Sensor Data Fusion*, 1989.

[101] K. Hwang and F.A. Briggs. *Computer architecture and parallel processing.* McGraw-Hill, New York, 1984.

[102] Y. K. Hwang and N. Ahuja. Path planning using a potential field representation. Technical Report UILU-ENG-88-2251, University of Illinois Coordinated Science Laboratory, 1988.

[103] Y. K. Hwang and N. Ahuja. A potential field approach to path planning. *IEEE Trans. Robotics and Automation*, 8(1), 1992.

[104] J. Ilari i Valentí, , and C. Torras i Gení s. 2D path planning: A configuration space heuristic approach. *Int. J. Robotics Res.*, 9(1):75–91, February 1990.

[105] M. Inaba and H. Inoue. Vision based robot programming. In *Proc. of 5th Intl. Symposium on Robotics Research*, 1989.

[106] Cybermation Inc. The K2A mobile platform. Sales literature, 1990. 5457 JAE Valley Road, Roanoke, VA 24014.

[107] P. Jacobs and J. F. Canny. Planning smooth paths for mobile robots. In *IEEE Int. Conf. Robotics and Automation*, pages 2–7, May 1989.

[108] P. Jacobs and J. F. Canny. Planning smooth paths for mobile robots. In Z. Li and J. F. Canny, editors, *Nonholonomic Motion Planning.* Kluwer Academic Publishers, 1993.

[109] P. Jacobs, J. P. Laumond, and M Taïx. Efficient motion planners for nonholonomic mobile robots. In *IEEE/RSJ Int. Conf. Intelligent Robot Systems*, Osaka, Japan, November 1991.

[110] G.M. Jenkins and D.G. Watts. *Spectral Analysis and its Applications.* Holden-Day Inc: San Francisco, 1968.

[111] G. Jones and M. Goldsmith. *Programming in Occam2.* Prentice Hall Int., 1988.

[112] L.P. Kaelbling. An architecture for intelligent reactive systems. In M.P. Georgeff and A.L. Lansky, editors, *Reasoning about actions and plans*, pages 395–410. Morgan-Kaufmann, 1987.

[113] Y. Kanayama and N. Miyake. Trajectory generation for mobile robots. In *Third Int. Symp. Robotics Research*, pages 333–340, Gouvieux, 1985.

[114] L. E Kay. Airborne ultrasonic imaging of a robot work space. *Sensor Review*, 1985.

[115] T. Keieriemen and E. von Puttkamer. Real-time control in an autonomous mobile robot. In G. Schmidt, editor, *Information Processing in Autonomous Mobile Robots*, pages 187–200. Springer-Verlag, 1991.

[116] O. Khatib. Real-time obstacle avoidance for manipulators and mobile robots. In *Int. Conf. Robotics & Automation*, pages 500–505, St. Louis, March 1985.

[117] O. Khatib. Real-time obstacle avoidance for manipulators and mobile robots. *Int. J. Rob. Research*, 5(1), 1986.

[118] C.A. Klein and W. Wahawisan. Use of a multiprocessor for control of a robotic system. *Int. J. Robotics Research*, 1(2):45–59, 1982.

[119] D. Koditschek. Exact robot navigation by means of potential functions: some topological considerations. In *IEEE Int. Conf. Robotics and Automation*, pages 1–6, Raleigh, North Carolina, 1987.

[120] Y. Koren and J. Borenstein. Potential field methods and their inherent limitations for mobile robot navigation. In *Proc. IEEE Int. Conf. Robotics and Automation*, pages 1398–1404, Sacramento, California, 1991.

[121] B. H. Krogh. A generalised potential field approach to obstacle avoidance control. In *Robotics Research: the next five years and beyond, SME Conference proceedings*, August 1984.

[122] B.H. Krogh and C.E. Thorpe. Integrated path planning and dynamic steering control for autonomous vehicles. In *Proc. IEEE Int. Conf. Robotics and Automation*, page 1664, 1986.

[123] D. T. Kuan, J. C. Zamiska, and R. A. Brooks. Natural decomposition of free space for path planning. In *IEEE Int. Conf. Robotics and Automation*, pages 168–173, St Louis, March 1985.

[124] R. Kuc and M.W. Siegel. Physically based simulation model for acoustic sensor robot navigation. *IEEE Transactions on Pattern Analysis and Machine Intelligence*, 9 6:766–778, 1987.

[125] Roman Kuc and Billur Barshan. Navigating vehicles through an unstructured environment using sonar. In *Proc. IEEE Int. Conf. Robotics and Automation*, pages 1422 - 1426, 1989.

[126] H. Kuttruff. *Ultrasonics: Fundamentals and Applications.* Elsevier Applied Science, 1991.

[127] S.Y. Lang, L.W. Korba, and A.K. Wong. Characterizing and modelling a sonar sensor. In *Proc. of SPIE*, pages 291–304, 1989.

[128] Jean-Claude Latombe. *Robot Motion Planning.* Kluwer Academic Publishers, 1991.

[129] J. P. Laumond, T. Siméon, R. Chatila, and G. Giralt. Trajectory planning and motion control for mobile robots. In J. D. Boissonnat and J. P. Laumond, editors, *Geometry and Robotics*, pages 133–149. Springer Verlag, 1989. Lecture notes in Computer Science, Vol 391.

[130] J.P. Laumond. Feasible trajectories for robots with kinematic and environmental constraints. In L. O. Hertzberger and F. C. A. Groen, editors, *Intelligent Autonomous Systems*, pages 346–354. North-Holland, 1986.

[131] D.T. Lawton, R.C. Arkin, and J.M. Cameron. Qualitative spatial understanding and reactive control for autonomous robots. In *IROS*, pages 709–714, Japan, 1990.

[132] J.J. Leonard. *Directed Sonar Sensing for Mobile Robot Navigation.* PhD thesis, University of Oxford, 1991.

[133] J.J. Leonard and H.F. Durrant-Whyte. *Directed Sonar Navigation.* Kluwer Academic Press, 1992.

[134] Inmos Limited. *OCCAM 2 Reference Manual.* Prentice Hall International (UK) Limited, Hertfordshire, U.K., 1988.

[135] Inmos Limited. *The Transputer Applications Notebook—Systems and Performance.* Redwood Burn Ltd., Trowbridge, U.K., 1989.

[136] Inmos Limited. *The Transputer Databook.* Redwood Burn Ltd., Trowbridge, U.K., 1989.

[137] F. R. Livingstone and M. Rioux. Development of a large field of view 3-D vision system. In *SPIE 665*, pages 188–194, 1986.

[138] A. Macoviski. Ultrasonic imaging array. *IEEE Proceedings*, 67 4:484–495, 1979.

[139] J. Manyika. *An Information-theoretic Approach to Data Fusion and Sensor Management.* PhD thesis, Oxford University, 1993.

[140] J.M. Manyika and H.F. Durrant-Whyte. An information-theoretic approach to management in decentralized data fusion. In *Proc. Spie 92 Conference. Vol.1828*, 1992.

[141] J.M. Manyika and H.F. Durrant-Whyte. Navigation using tracking sonars. Technical Report 1956/92, Oxford U. Robotics Research Group, 1992.

[142] G.L. Miller. and E.R. Wagner. An optical rangefinder for autonomous robot cart navigation. In *SPIE 852*, pages 122–134, 1987.

[143] J. Minkoff. *Signals, Noise and Active Sensors.* AT and T, 1992.

[144] T.M. Mitchell. Becoming increasingly reactive. In *Proc. 8th National Conf. on Artificial Intelligence*, pages 1051–1058, 1990.

[145] H. Moravec. Sensor fusion in certainty grids for mobile robots. In *Sensor Devices and Systems for Robotics*, pages 253–276. Springer-Velag. Nato ASI Series, 1989.

[146] H.P. Moravec and A. Elfes. High resolution maps from wide angle sonar. In *Proc. IEEE Int. Conf. Robotics and Automation*, pages 116–121, 1985.

[147] P.M. Morse and K.U. Ingard. *Theoretical Acoustics.* New York: McGraw-Hill, 1968.

[148] W.S. Munro, S.C. Pomeroy, M. Rafiq, H.R. Williams, M.D.Wybrow, and C. Wykes. Ultrasonic vehicle guidance transducer. *Ultrasonics*, 28:349–354, 1990.

[149] N. Nilsson. Shakey the robot. Technical Report 323, SRI, April 1984.

[150] A. Noble and M. Brady. Estimating asymmetric noise. In *Proc. of the 5th Alvey Vision Conference*, pages 157–162, University of Reading, U.K., September 1989.

[151] F.R. Noriels and R.G. Chatila. Control of mobile robot actions. In *Proc. IEEE Int. Conf. Robotics and Automation*, pages 701–707, 1989.

[152] Colm Ó'Dúnlaing and Chee K. Yap. A retraction method for planning the motion of a disc. In J. T. Schwartz, M Sharir, and J Hopcroft, editors, *Planning, Geometry, and Complexity of Robot Motion.* Anlex Publishing Corporation, 1987.

[153] G.L. Oomen and W.J.P.A. Verbeek. A real-time optical profile sensor for robotic arc welding. In *SPIE 449*, pages 62–71, 1984.

[154] A. Oppenheim and R. Schafer. *Digital Signal Processing.* Englewood Cliffs, NJ: Prentice-Hall, 1975.

[155] N.E. Orlando. An intelligent robotics control scheme. In *American Control Conference*, page 204, 1984.

[156] R.P. Paul, H.F. Durrant-Whyte, and Max Mintz. A robust distributed robot control system. In *Third Int. Symp. Robotics Research*, France, 1985.

[157] D.W Payton. An architecture for reflexive autonomous vehicle control. In *Proceedings of IEEE Conference of Robotics and Automation*, page 1838, 1986.

[158] N.E. Pears and P.J. Probert. Active triangulation rangefinder design for mobile robots. In *Proc. IEEE/RSJ Int. Conf. on Intelligent Robots and Systems*, pages 2047–2052, 1992.

[159] N.E. Pears and P.J. Probert. A lateral effect photodiode based scanning sensor for autonomous vehicle obstacle avoidance. In *1st IFAC Int. Workshop on Intelligent Autonomous Vehicles*, 1993.

[160] N.E. Pears and P.J. Probert. An optical range sensor for mobile robot guidance. In *Proc. IEEE Int. Conf. on Robotics and Automation*, pages 3.659–3.665, 1993.

[161] C. M. Pearson and P. J. Probert. A monocular feature tracker for low cost stereo vision control of an AGV. In *SPIE International Symposium on Optical Tools for Manufacturing and Advanced Automation, Mobile Robotics*, 1993.

[162] Jian Peng and Stephen Cameron. Experiments with a rule-based robot palletiser. In Spyros G. Tzafestas, editor, *Engineering Systems with Intelligence*, pages 117–124, Corfu, June 1991. Kluwer Academic Publishers. Proceedings EURISCON-91.

[163] F. G. Pin and H. A. Vasseur. Autonomous trajectory generation for mobile robots with non-holonomic and steering angle constraints. In *IEEE International Workshop on Intelligent Motion Control*, Istanbul, August 1990.

[164] J. K. Poon and P. D. Lawrence. Path planning with a potential surface in configuration space. Department of Electrical Engineering, University of British Columbia, 1988.

[165] Dick Pountain and David May. *A Tutorial Introduction to Occam Programming*. BSP Professional Books, 1988.

[166] F. P. Preparata and M. I. Shamos. *Computational Geometry*. Springer-Verlag, New York, 1985.

[167] P.J. Probert and H. Hu. Transputer architectures for mobile robots. In *IEE Colloquium on Robot System Architectures*, 1990.

[168] J. A. Reeds and R. A. Shepp. Optimal paths for a car that goes both forward and backwards. *Pacific Journal of Mathematics*, 145(2), 1990.

[169] U. Regensburer and V. Graefe. Object classification for obstacle avoidance. In *Proc. SPIE Symposium on Advances in Intelligent Systems*, 1990.

[170] G. T. Reid, S. J. Marshall, R. C. Rixon, and H. Stewart. A laser scanning camera for range data acquisition. *J. Phys. D: Appl. Phys.*, 21:1–3, 1988.

[171] I. D. Reid. Recognition of parametrised models from 3D data. In *Proceedings of the IAPR Workshop on Machine Vision Applications*, pages 347–350, 1990.

[172] I. D. Reid and J. M. Brady. Recognizing parameterized models using 3D edges. In P. Mowforth, editor, *Proceedings of the British Machine Vision Conference*, pages 355–358. Springer-Verlag, 1991.

[173] I. D. Reid and J. M. Brady. Recognition of object classes from range data. In *Int. Conference on Computer Vision*, 1993.

[174] I. D. Reid, J. M. Brady, A. McIvor, S. Marshall, I. B. Knight, and R. C. Rixon. Range vision and the Oxford AGV. In *Proc. Image Processing '89, London*, pages 51–72. Blenheim Online Publications, October 1989.

[175] Ian Reid and Michael Brady. Model-based recognition and range imaging for a guided vehicle. *Image and Vision Computing*, 10(3), April 1992. Special issue on Range Image Understanding.

[176] E. Rimon and D. E. Koditschek. Exact robot navigation in geometrically complicated but topologically simple spaces. In *IEEE Int. Conf. Robotics and Automation*, pages 1937–1942, Cincinnati, May 1990.

[177] M. Rioux. Laser rangefinder based on synchronised scanners. *Applied Optics*, 23(21):3837–3844, 1984.

[178] M.E. Rozmann and J. Detlefsen. Environmental exploration based on a three dimensional imaging radar sensor. In *RSJ/IEEE International Workshop on Intelligent Robots and Systems*, pages 422–429, 1992.

[179] Hanan Samet. The quadtree and related hierachial data structures. *ACM Computing Surveys*, 16(2):187–260, June 1984.

[180] K.F. Sander and G.A.L. Reed. *Transmission and Propagation of Electromagnetic Waves.* Cambridge University Press, 1978.

[181] V. Sarkar. *Partitioning and scheduling parallel programs for multiprocessors.* Pitman publishing, 1989.

[182] J. F. Silverman and D. B. Cooper. Bayesian clustering for unsupervised estimation of surface and texture models. *IEEE Transactions on Pattern Analysis and Machine Intelligence*, 10(4):482–495, 1988.

[183] S. Singh and P. Keller. Obstacle detection for high speed autonomous navigation. In *IEEE Int. Conf. on Robotics and Automation*, pages 2798–2803, 1991.

[184] S. Singh and M. D. Wagh. Robot path planning using intersecting convex shapes. In *IEEE Int. Conf. Robotics and Automation*, pages 1743–1748, San Francisco, April 1986.

[185] M.I. Skolnik. *Introduction to Radar Systems.* McGraw Hill Intl., 1981.

[186] M. Stephens and C. Harris. 3D wire-frame integration from image sequences. In *Alvey Vision Conference*, Sept. 1988.

[187] O. Takahashi and R.J. Schilling. Motion planning in a plane using generalised Voronoi diagrams. *IEEE J. Robotics and Automation*, RA-5 2, 1989.

[188] R. B. Tilove. Local obstacle avoidance for mobile robots based on the method of artificial potentials. In *IEEE Int. Conf. Robotics and Automation*, pages 566–571, Cincinatti, May 1990.

[189] P. Tournassoud and O. Jehl. Motion planning for a mobile robot with a kinematic constraint. In *IEEE Int. Conf. Robotics and Automation*, pages 1785–1790, Philadelphia, April 1988.

[190] P. Tournassoud and O. Jehl. Motion planning for a mobile robot with a kinematic constraint. In J. D. Boissonnat and J. P. Laumond, editors, *Geometry and Robotics*, pages 150–171. Springer Verlag, 1989. Lecture notes in Computer Science, Vol 391.

[191] S.A. Walter. The sonar ring: Obstacle detection for a mobile robot. In *Proc. IEEE Int. Conf. Robotics and Automation*, pages 1574–1579, 1987.

[192] H Wang and C C Bowman. The Oxford distributed machine for 3D vision system. In *IEE colloquium on Parallel Architectures for Image Processing Applications*, pages 1/2–5/2, London, April 1991.

[193] H Wang, P M Dew, and J A Webb. Implementation of Apply on a transputer array. *CONCURRENCY: Practice and Experience*, 3(1):43–54, February 1991.

[194] Han Wang and Mike Brady. Corner detection for 3D vision using array processors. In *BARNAIMAGE 91*, Barcelona, Sept. 1991. Springer-Verlag.

[195] Wu Wen. *Multi-Sensor Geometric Estimation.* PhD thesis, Oxford University, 1992.

[196] G. T. Wilfong. Motion planning for an autonomous vehicle. In *IEEE Int. Conf. Robotics and Automation*, pages 529–533, Philadelphia, April 1988.

[197] N. Yokoya and M. D. Levine. Range image segmentation based on differential geometry: a hybrid approach. *IEEE Transactions on Pattern Analysis and Machine Intelligence*, 11(6):643–649, 1989.

[198] S. Yuta, S. Suzuki, and S. Iida. Implementation of a small size experimental self-contained autonomous robot. In *Second International Symposium on Experimental Robotics*, Toulouse, June 1991.

[199] Y. Zhao and S.L. BeMent. A heuristic search approach for mobile robot trap recovery. In *Proc. SPIE*, volume 1388, Mobile Robot V, pages 122–130, 1990.

Index